THE EFFECTIVE USE OF STATISTICS

A Practical Guide for Managers

TIM HANNAGAN

In Memory of Silver Birches

First published in 1990 by
Kogan Page Ltd,
120 Pentonville Rd, London N1 9JN
in association with The Chartered Institute of Management Accountants,
63 Portland Place, London W1N 4AB.

Printed and Bound in Great Britain by Biddles Ltd, Guildford & Kings Lynn.

British Library Cataloguing in Publication Data
Hannagan, T. J.
The effective use of statistics.
1. Statistical mathematics
I. Title
519.5

ISBN 1-85091-911-9

Contents

List of Figures

List of Tables

Preface

The aim of this book is to provide a statistical foundation for managers, some of whom may have very limited mathematical knowledge. The book is designed to enable managers to interpret and present effectively the statistical information which is a day-to-day part of their work. The book is organised to give a comprehensive understanding of statistical sources, concepts, methods and applications used in management and to be a handy reference for managers. The objective of the book is to stress the relevance of statistics to managers in industry, commerce, service industries, the public sector, public services and non-profit organisations.

The book is designed for all managers who need to understand, interpret and make use of statistical information in the course of their work. It will provide a reference and handbook for managers who need to remind themselves of a particular technique, who want to gather information or who want ideas about presenting reports.

A section on writing management reports provides help and advice on writing reports and making presentations. The book is full of examples of the application of statistical information in management and the use of statistics for managers to reinforce the ideas and information contained in the text. A list of formulae and a glossary of the main statistical terms are included for easy reference as 'bullet points' at the end of each chapter.

On completion of this book the reader should be aware and have an understanding of the use of statistical information in management, the main sources of this information and its interpretation, and be proficient in the basic statistical methods and forms of presentation used in management.

Tim Hannagan
Chalfont St Giles
October 1989

Acknowledgements

The author and publishers wish to thank the Controller of Her Majesty's Stationery Office for permission to use tables from *Economic Progress Reports* (February and April 1989) and the Registrar General of the Office of Population Censuses and Surveys for permission to use tables from the *Population Census* (1981).

The author wishes to thank Alison Rollins for her ability to translate his writing into typescript.

Part 1
Statistical Information for Managers

1
The Manager and Statistics

The nature of management

A manager can be described as a person who decides what needs to be done and who arranges for someone else to do it. This suggests that the management role involves decision-making, choice, supervision and control rather than a purely operational function. Few airline managers continue to fly, for example, and a sales manager has a team of salesmen actually to sell the product or service. If the sales manager does sell directly he is acting as one of the salesmen rather than as a manager.

Management includes the need to forecast and plan, to organise, to command and co-ordinate. The role does not mean following instructions or carrying out routine tasks as is the case in an operational job; it does mean making decisions between different courses of action. A managerial job offers opportunities for making choices in what is done, how it is done and when it is done. It can be argued that most forms of work, with the possible exception of the most routine or automatic, imply making some choices so that most work has an element of 'management' in it and it is important to identify this element and the proportion it makes up of the total job.

It is perhaps possible to 'measure' a management job by the length of time the manager is left to work on his or her own before the work is checked for quality. This 'time-space discretion' may be a few hours to a few days for people working at the lower levels in an organisation and much longer for those at the top of the organisation. It would seem to be true also that the more choice a manager has, the further up the management hierarchy he or she has climbed. A junior manager has a limited area of choice and limited room to manoeuvre but the management function can be said to have clearly entered a job at the moment when a significant amount of decision-making and choice enter the role.

Promotion in an organisation normally means an increase in the management function, so that decision-making and choice become an increasing part of the job. The move from the role of 'player' to that of 'manager' requires considerable adjustment and often is an irrevocable break which may be difficult for people to make. The brilliant salesman for example may not be a good sales manager. The outstanding teacher may make a poor head teacher. Top managers are concerned much more with planning and organising and will spend relatively little time supervising, while middle managers may spend fairly even amounts of time on planning, organising, directing and controlling. Figure 1.1 shows how the development of the management role can be represented diagrammatically.

It would be wrong to confuse management and ownership which may go hand in hand but are often separated. To manage is to have effective control of a particular area of work; a football manager, for example, has effective control of the team even though he does not own the club. All the functions which have been identified as being part of management, such as organising, controlling, decision-making, planning and forecasting, involve the use and application of figures and it is very important therefore for managers to understand the nature of statistics.

The nature of statistics

Statistics is concerned with scientific methods for collecting, organising, summarising, presenting and analysing data as well as drawing valid conclusions and making reasonable decisions on the basis of this analysis. Data (singular 'datum') are things known or granted as a basis for inference. Statistics is concerned with the systematic collection of numerical data and its interpretation.

The word 'statistics' is used to refer to:

i numerical data, such as the number of people employed in an organisation;

ii the study of ways of collecting and interpreting these facts.

It can be argued that figures are not facts in themselves; it is only when they are interpreted that they become relevant to discussions and decisions.

People receive large quantities of information every day through conversations, television, the radio, newspapers, posters, notices and instructions. Managers are inundated with information from government reports, official statistics, company memos, legal documents, financial returns, advertising literature, trade journals and many other sources.

It is just because there is so much information available that

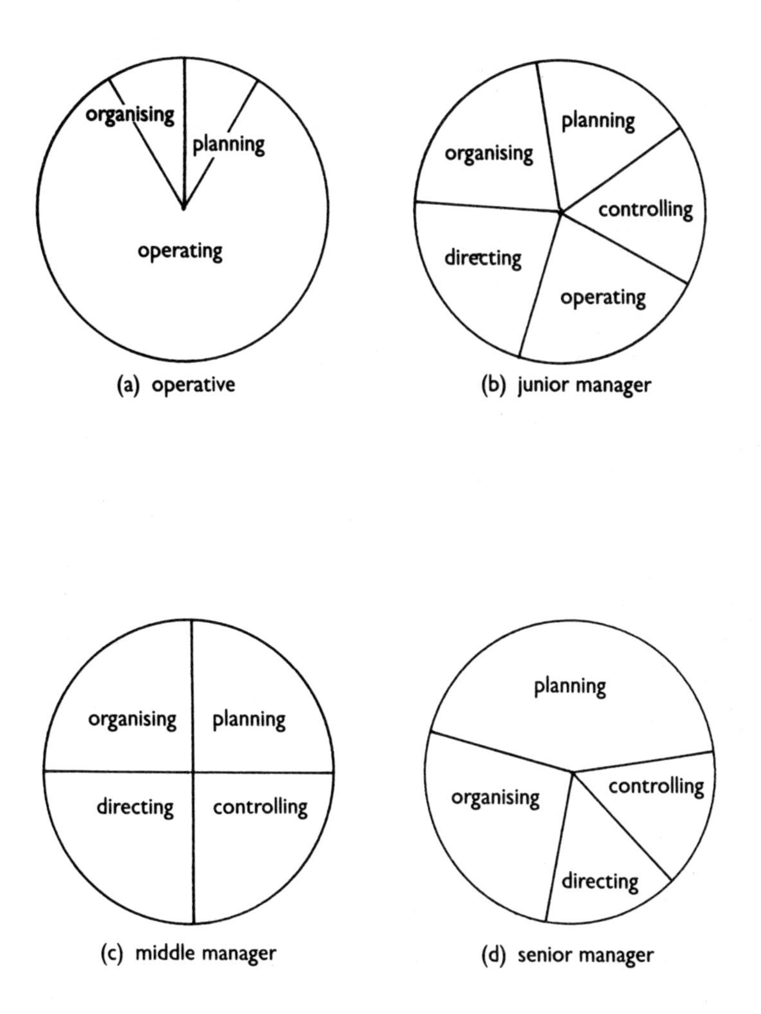

Figure 1.1 *The managerial role*

people need to be able to absorb, select and reject it, to sift through facts to pick out what is interesting and useful. In everyday life as well as in business and industry, certain information is necessary and it is important to know where to find it or how to collect it. As consumers, everybody has to compare prices and quality before making a decision about what goods to buy. As employees, people want to compare wages and conditions of work. As employers, firms want to control costs and expand profit. Everybody collects, interprets and uses information, much of it in a numerical or statistical form.

The ease of processing numerical data through the development of information technology has increased their use and availability; however, there is a tendency to avoid numerical information in favour of written, visual, or verbal information. Few newspapers publish any significant amount of statistics and the media in general tend to ignore the range and variety of statistical publications available. There is a tendency to translate statistics into a verbal form. The headline, 'Low Unemployment and Consumer Boom continues', is a statistical statement. It is an interpretation of data on the levels of employment and data on retail sales figures. Such words as 'greater', 'more', 'higher', 'lower', 'small', 'insignificant', all imply quantification. So in fact do words like 'boom', 'slump', 'increase', 'decrease', 'rise', 'fall', 'profit', 'insolvency' and so on. To say or write that a company has an increasing profit or is heading for insolvency may be summaries of complex sets of figures.

One of the interesting exceptions to this avoidance of statistics is in sport, where complicated tables and figures are accepted as normal. People have little difficulty in understanding football league tables, batting averages or betting odds. The Cockcroft Report on Mathematics stated that: 'It would be very difficult—perhaps impossible—to live a normal life in very many parts of the world in the twentieth century without making use of mathematics of some kind.' Perhaps the ability of so many people to work with sports data indicates that everybody is capable of understanding complicated numerical information if they are sufficiently well motivated.

It is interesting also to notice that to be illiterate is considered unacceptable but to be innumerate is not. People say that they are 'no good with figures' and look at tables and graphs without comprehension. Managers cannot afford to be in this position. They do not have to be accountants or statisticians but they do need to be able to understand basic forms of statistical presentation and analysis. In fact most managers, like most people, have more facility with figures than they often fully realise. In the same way that many facts about sport can be provided most clearly in numerical form, so most facts about money include elements of numeracy. People know how much change they should receive to the nearest penny, how much they earn to the nearest one, ten or hundred pounds and they have a reasonable idea of what happens to their money when there is rapid inflation.

As the numerical information moves further away from personal detail towards more general data, problems may arise in interpretation. In negotiating pay rises there is often dispute about what the figures mean. A 5 per cent pay rise for everyone in a company may appear simple and straightforward but in fact has a range of implications. Those employees earning the most will receive the highest increase (5 per cent of £10,000 is £500; 5 per cent of £20,000 is £1,000); if the overall cost of living has risen by 10 per cent since the last pay rise the rise will not appear to be generous; employees may work in different branches of the company and one area may have lower living costs than another, so that a rise is more valuable in one area than another. Often pay negotiations arrive at complex agreements just because of these factors.

This illustrates the point that numerical data on their own do not provide any information at all. The statement that 'there has been a 5 per cent pay rise in this company' does not have any significance unless it is set against such factors as inflation and when the last pay rise was agreed. Interpretation is a matter of judgement based on knowledge of what is meant by the term 'inflation'. An important point which is not always sufficiently emphasised in school mathematics is that it is at least as important that a figure or the result of a calculation is understood as it is that it is accurate. There is little point in arriving at a correct answer to a calculation if it is not known what it means.

The use of statistics in management

Information is an essential element in management and much of this is in statistical form. Managers need statistics to help set targets and objectives to monitor performance against standards, to exercise control and to assist in making decisions. Everybody uses statistics to compare prices in shops, to look at the performance of new cars, to buy a house and so on; managers use statistics in similar ways and for a range of activities, from buying policy to quality control, market research, investment decisions, forward planning and union bargaining.

A complex society would have difficulty in functioning without statistics. Government actions are more likely to be effective if they are based on good information. Governments use statistics in making policy decisions and monitoring their results and in trying to understand economic and social relationships. A modern industrial economy is so complex and interdependent that it is necessary to know a great deal about its various facets in order to understand what is happening and decide upon the policies needed and what their effects might be. If manufacturing output seems to have started to fall, for example, there is a need to know what is happening to stocks, to order books, to prices and to employment, whether the fall is due to consumers spending less, to falling exports, rising imports, or a decline in capital spending.

Large companies have similar needs and these are the basis for the production of many statistics published by government departments and agencies. They also need statistics for their own internal management purposes. Statistics are a vital part of decision-making because they can narrow the area of disagreement and can help to define a problem once it has been recognised to exist. A large retail store may sense that it has a car parking problem, for example, which is causing aggravation to its customers. The dimensions and nature of the problem can be discovered by collecting information on the number of spaces available against the number of cars trying to park at various times of day on different days of the week. Further information on the length of queues for the car park, the number of cars driving away over a specific time period, and so on, may help as well. All this statistical information helps to define the problem and narrow any area of disagreement about its nature.

Statistics in management can be a stabilising force, dispelling rumour and uncertainty, helping to solve arguments arising from individual cases or circumstances by providing a factual foundation to debates and decisions which would otherwise be dominated by subjectively based theories and opinions. At its simplest, a discussion about who won the football match last night is decided by the news that Team A won by 2 goals to nil.

This accurate statistical information ends that argument but not of course the whole discussion about who should have won or the way in which Team A won and Team B lost. In more complex management terms the sales manager of a company may feel that the sales of a product have fallen because his sales force is not working hard enough; marketing research may show that in fact the market for the product has shrunk so that all brands of the product have experienced falling sales and that the sales manager's company has actually increased market share.

In a sense statistics are the opposite of the 'Grandmother Theory'. This theory runs on the lines of 'my Grandmother (or Grandfather) smoked 50 cigarettes (drank a bottle of whisky/ate like a horse) every day and lived to be 100 years old; therefore smoking (drinking, over-eating) cannot be bad for you.' Statistics do not prove that everybody who smokes heavily dies young, but the statistical evidence does show that people are more likely to die young if they smoke heavily than if they do not.

It is this feature of systematic collection of data that distinguishes statistics from other kinds of information and makes it of particular value to management. Managers need information about what has happened in the past, what is happening now and what is expected to happen in the future. Statistics are an aid to the exercise of management control. Managers need statistical information to help them to set targets and objectives, to monitor performance against set standards and in deciding what to do when the performance has been compared with these standards, and they need to know if objectives have been achieved and if the best results have been obtained from the available money, people and equipment. Managers need to understand statistical data, to be able to collect statistical information if it is not available already and to apply statistical techniques to their work, and they need to be able to quantify and summarise business-related data from a range of sources, to tackle business problems requiring statistical analysis and to present reports and recommendations arising from the analysis of statistical data. Statistics provide a method of summarising in a systematic way aspects of the complexities of business and economic problems, and managers use:

descriptive statistics: which include the presentation of data in tables and diagrams as well as calculating management ratios, percentages, averages, measures of dispersion and correlation, in order to display the salient features of the data and to reduce them to manageable proportions, and:

inductive statistics: which involves methods of inferring properties of a population (total set of items under consideration) on the basis of known sample results (based directly on probability theory).

In fact most knowledge is 'probability knowledge' in the sense that it is possible to be absolutely certain that a statement is true only if it is of a restricted kind. Statements such as: 'I was born after my father', 'a black cat is black', 'one plus one equals two', are tautologies or disguised definitions, and although they are true they are of limited value in providing information. Much information is based on sample data rather than on a complete survey of a population and probability theory is the basis of the areas of statistics that are not purely descriptive. Managers tend to be largely concerned with descriptive statistics when they are working with financial returns, wage lists, personnel details, production figures and management ratios. They become involved with inductive statistics when they are working with samples and statistical decisions in such aspects of their work as marketing and market research.

Management information systems with the use of computers have had a major impact on the sources of data available to managers. Routine reports can be produced more rapidly than in the past so that the report can be received only a short time after the activity to which it refers has taken place. Information can be collated from a number of different sources and individual managers can have their own processing power, for example, a micro-computer, so that they can generate their own statistics. They may use such methods as desktop publishing to produce well-presented reports incorporating tables and diagrams.

The sources of information for managers

Management information comes from two major sources: information collected by managers and their organisations for themselves, and information collected by other people. In statistical terms we have:

i *Primary data*: statistical information collected by managers or their organisations.

ii *Secondary data*: statistical information collected by somebody else such as a government department, another company or a research organisation.

The collection of *primary data* involves all the survey and sample methods available to managers which are described in Chapter 3. Once these data have been collected, processed and published they are available for use by other people as secondary data. *Secondary data* are in this sense 'second-hand'. Companies, government departments and public sector organisations generate large quantities of data for a variety of reasons. When they use them themselves these are primary data, when they are used by other people they are described as secondary data. Managers will use large quantities of data produced by others and because these are second-hand it is important, just as it is with a second-hand car, to know as much as possible about them. It is useful to know how the data have been collected and processed in order to appreciate their reliability and to understand the full meaning of the statistics.

Managers need to ask:

i How have the data been collected?
ii How have the data been processed?
iii How accurate are the data?
iv How have the data been summarised?
v How comparable are the data with other tabulations?
vi How should the data be interpreted?

In spite of the care required in using secondary data they will still be worth using if they are of sufficient quality. It may not be known exactly 'where they've been' or 'how they have been put together' but after testing, they may be found to be of sufficiently good quality to use.

There are great advantages in using secondary data:

i *Cost*: Government publications are relatively cheap and libraries stock quantities of data produced by the government, by companies and by other organisations.
ii *Availability*: They are readily available in large quantities and there is a great variety of data on a wide range of subjects. Much secondary data have been collected for many years and therefore can be used to plot trends.
iii *Use*: Secondary data may be useful to managers in areas such as marketing and sales, in order to appreciate the general economic and social condition, and they provide information on competitors.

The main problem with using secondary data in management

is that they are unlikely to be exactly what is wanted. The data have been produced for a particular purpose, often as part of an 'administrative process' or a special survey and they are unlikely to be exactly what is wanted for another purpose. Managers themselves will generate wage lists, production and cost figures which may be of little use as they stand to anybody outside the company. Other people's administrative data may be equally of little interest outside their organisation. Even when they are of interest, statistics collected as part of an administrative process may not be entirely appropriate for other uses.

Statistics on unemployment, for example, are of general interest but the data are collected by an administrative process which means that they may be of more or less use to particular managers. The statistics are collected mainly as part of the process of registration by individuals as unemployed in order to be eligible for unemployment benefit. Large numbers of people who are looking for work do not in fact register, either because they expect to find work at any moment and do not bother to register, or because they are not eligible for benefit (housewives, for example). A manager who wants to have an accurate picture of the labour supply in a particular area may use unemployment statistics as an estimate but needs to carry out further investigations.

The statistical information which is widely available and of use to managers can be divided into:

i *Micro-statistical information*: which is data produced by private firms and private organisations. It is information specific to the business organisation or the public sector institution, compiled in the process of monitoring activities of the business or public sector unit. It may include information about aspects of production, costing, marketing, labour supply and so on.

ii *Macro-statistical information*: data produced by the government or government department or private research body or a public sector organisation and which relate to the country as a whole rather than to one organisation or unit. These data include information about population, finance, education, unemployment, social changes and so on. Government departments, employers' federations, trade associations, trades unions, private firms, professional institutes, and public and private research organisations collect and publish large quantities of statistical information.

As managers move away from an operational base to more general management they may need to have access to macro-statistical information to set their activities in context. Marketing managers, for example, will need to consider the economic and social environment in which they are working as part of their marketing audit.

The abuse of statistics

Statistics are numerical facts and as such they are a tool to be used in management as a spade might be used in gardening. A spade is designed for digging but may be used for hammering in a post. When it is used in this way it is being abused, so if it breaks the fault lies with the user, not the manufacturer of the spade. Similarly, statistics can be abused, usually by interpretation or misinterpretation, and it is not the statistics which are at fault, it is the interpreter.

Managers need to be aware that: 'statistics can be used like a drunken man uses a lamp post, for support rather than illumination.' Statistics are often manipulated to support an argument and figures may be 'massaged' in order to support opinions or prejudices. If the objective is decided and then there is a search for figures to support it, the likelihood is that this search will be successful. A number of octogenarians can be found who have smoked all their lives but this is not statistical proof of the benefits of smoking.

In an ideal situation managers will collect and analyse statistics as accurately and objectively as possible and will expect the same to have been done with the data collected for them. If there is an argument about levels of productivity in a group of workers, it is first necessary to agree on what is meant by 'productivity' and how it can be measured, and then the data need to be collected and interpreted without prejudice. This is very difficult for managers, who may have staked their reputation on a particular outcome, so that it is much better to collect the information at an early stage, and before a conclusion is needed, in order to narrow the argument.

Once managers have decided 'to look at the facts', abuse can still take place at each stage. Abuse can arise from decisions about what numerical facts to collect, how to collect them, when and where and then how to classify and analyse them. If statistics are collected or interpreted to hammer in an argument, like a spade to hammer in a post, they are not being used correctly. They should provide a factual base for discussion and decision-making in order to help to remove areas of disagreement.

- *Managers* makes decisions and choices, and plan and control.
- *Statistics* is concerned with scientific methods for collecting, organising, summarising, presenting and analysing data.
- *Managers use statistics* to help set targets and objectives, to monitor standards, exercise control, to assist in making decisions, and to reduce the area of disagreement about the nature of a problem.
- *Primary data* are collected by managers or their organisations. *Secondary data* are collected by somebody else. They are cheap and available but not always exactly what a manager wants.
- Managers need to ask *how* data have been collected, processed, summarised, tabulated and interpreted.
- *Sources of data* include firms, private organisations, public institutions, the government and its departments. Managers can use information from such sources in their own production, marketing, sales, finance and personnel departments as well as in planning and forecasting.
- Statistics can be *abused* and used more for support than illumination.

2
Effective Presentation for Managers

Effective communication

The use of presentation

Effective presentation is the skill of using the best format for a particular audience at a particular time. A slogan may be the best method for communicating a simple idea to many people, while a detailed report may be the best format for the analysis of a specialised subject.

The purpose of communication is both to inform people and also to persuade them. The impact is important if facts and ideas are not to be overlooked. In advertising, the way a product or service is presented may sometimes appear to be more important than the facts, but if the message is not communicated the advertisement will have failed. In a wage dispute a banner reading 10 per cent may 'say it all'. In terms of establishing a position, however, the negotiators are likely to use more detailed arguments and statistical information.

Presentation is important both for imparting information and for providing arguments, analysis and persuasion. Verbal and written reports may be supported by tables, graphs and diagrams as methods of presenting and summarising statistical data.

Forms of presentation

The form of presentation used will depend on:

- i the subject matter
- ii the purpose of the communication
- iii the best method of imparting the message
- iv the best method of persuasion
- v the amount of detail required.

Very often graphs and diagrams will prove to be useful in presenting and illustrating information. Assertiveness training suggests that it is important to deliver a message in such a way that it is understood. It is not effective to be misunderstood, so that saying 'no' to the doorstep salesman requires a different approach and emphasis from saying 'no' to an offer of help at work. Both answers may include a 'thank you' and a reason for refusal, but the first is in order to be rid of the salesman expeditiously, while the second is aimed at maintaining the co-operation of a colleague.

A manager may have five minutes in which to present a case to the board of directors on a complex matter of policy and a few well chosen tables and graphs may make a greater impact than a purely verbal presentation. Long lists of figures will put off many audiences while a form of pictorial or diagrammatic summary will not. In some cases the most appropriate form of presentation will be a written report, in others it will be a series of tables, graphs and diagrams. Often it will be a combination of these methods using audio-visual equipment to focus attention on the essential features.

Statistical tables in management

Raw data

Raw data are often in the form of lists, such as wage lists, production schedules, and inventories. The result of surveys and questionnaires may also be listed with the number of people answering 'yes/no/don't know' to particular questions. Before these data can be used easily they need to be put in a form in which they can be interpreted, and the first step is to construct a table.

Tables are used to present figures in an orderly manner; they may help to summarise information and to show distinct patterns in the data. There are numerous examples of them in company reports, financial statements, sports results and government publications. Computers can produce tables in large volumes if required, which may involve weeks of analysis and interpretation and it is, therefore, important to decide on the aims and objectives of creating these tables.

There are a number of common features to look for in tables, and to include when constructing one:

- i All tables should have a title which indicates the content.
- ii Column and row headings should be brief and self-explanatory.
- iii The source of data should be included so that original

sources can be checked. This information is usually given immediately below the table.

iv Units of measurement should be shown.

v Sets of data which are to be compared should be close together.

vi Derived statistics, such as percentages, should be beside the figures to which they relate.

vii Approximations and omissions can be explained in footnotes.

Table 2.1 is a good example of this form of presentation.

Table 2.1 *International comparison of saving rates*[1]

		Private saving[2]	*National saving*
USA	1986	16.1	12.7
	1987	14.7	12.4
	1988[3]	15.9	13.9
Japan	1986	27.8	31.8
	1987	—	—
	1988[3]	—	—
Germany	1986	—	23.8
	1987	—	23.7
	1988[3]	—	25.3
France	1986	19.3	19.8
	1987	18.6	19.6
	1988[3]	18.7	21.1
UK	1986	18.9	18.3
	1987	19.7	19.8
	1988[4]	17.5	19.2
Italy	1986	28.4	22
	1987	—	—
	1988	—	—
Canada	1986	22.6	18.1
	1987	22.4	18.7
	1988[3]	22.7	20.2

—Not currently available on OECD databases.

[1]Ratio of saving (gross of capital consumption but net of stock appreciation) to GDP/GNP (%).

[2]Including public enterprises.

[3]1988 first half.

[4]1988 first quarter.

Source: OECD quarterly national accounts database—except for Italy: OECD annual national accounts database. Individual country data only partially adjusted to standardised definition.

[*Treasury Economic Progress Report*, February 1989]

Classification

This is the process of relating the separate items within the mass of data collected and the definition of various categories. It is a process which has to be carried out before data can be tabulated.

All data have characteristics, some of which are measurable and quantifiable, some of which are non-measurable attributes and are qualitative.

Qualitative data
Qualitative characteristics can be ranked on a subjective basis to produce results such as those seen in beauty contests, wine tasting, shades of colour and ideas about design. This ranking is a matter of judgement and although measurements may be involved these are based on subjective criteria.

Quantitative data
Measurable characteristics include such variables as length, weight and height.
Discrete variables are measured in single units (such as people, houses, cars) and only appear in a fractional format when they are averaged. Such statements as 'the average household in the south-east of England owns 1.2 cars' are possible because they refer to averages even though it may be difficult to define 0.2 of a car.
Continuous variables are in units of measurement which can be broken down into definite gradations, such as length in centimetres or inches, or temperatures in degrees and decimals or fractions of a degree. In producing a table decisions have to be made about gradations and class intervals.

Class intervals

Class intervals need to be shown clearly and unambiguously. With a discrete variable there does not need to be confusion between class intervals:

Number of people
10–19
20–29
30–39
40–49

It is clear that the 19th person will be in the first class and the

20th person in the second class. The position of the 20th person would be less obvious if the class intervals were shown as:

Number of people
10–20
20–30
30–40
40–50

With continuous variables the level of approximation or rounding needs to be decided at an early stage in tabulation. The length of planks of wood may be measured to the nearest centimetre:

Length (in metres)
3.20 m–3.59 m
3.60 m–3.99 m
4.00 m–4.39 m
4.40 m–4.79 m

The 'rules' of approximation and rounding will need to be followed here so that a plank which is 3 metres 59.4 centimetres will be in the first class, a plank at 3 metres 59.5 centimetres in the second.

Another way of expressing this is:

Length (in metres)
3.20 m but less than 3.60 m
3.60 m but less than 4.00 m
4.00 m but less than 4.40 m
4.40 m but less than 4.80 m

In this case decisions have to be made and noted as to whether 'but less than 3.60 m' means up to 3 metres 59.5 centimetres or 3 metres 59.99 centimetres.

Some tables have open-ended class intervals, normally at the beginning or end of a distribution. A wages and salary list, for example, may show employees in class intervals of twenty years:

Age of employees	*Average weekly pay (£)*
Up to 20	100
20 but under 40	160
40 but under 60	240
60 and above	220

The interval size of open-ended classes needs to be decided for

Constructing a statistical table

In preparing a fact-finding report, the number of orders telephoned into five branches of a company are noted and the number of those that become firm orders. A simple report form is used to produce the main data:

Branch	Telephoned orders (confirmed orders are circled)
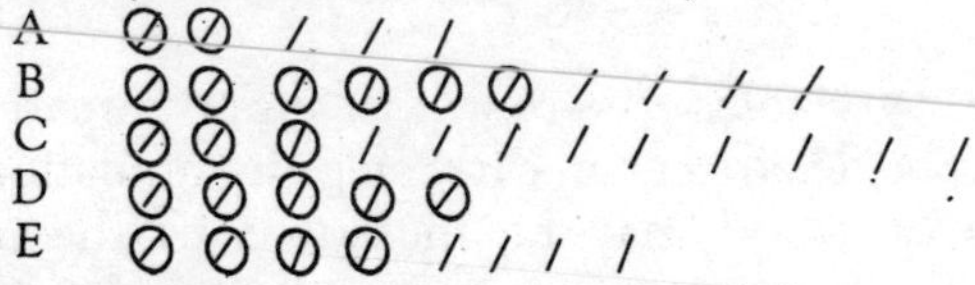 A	⊘ ⊘ / / /
B	⊘ ⊘ ⊘ ⊘ ⊘ ⊘ / / / /
C	⊘ ⊘ ⊘ / / / / / / / / /
D	⊘ ⊘ ⊘ ⊘ ⊘
E	⊘ ⊘ ⊘ ⊘ / / / /

A simple table can be constructed from this:

Table 2.2 *Telephoned and confirmed orders to branches of company X*

Branch	*Telephoned orders (number)*	*Confirmed orders (number)*	*Percentage of confirmed to telephoned orders*
A	5	2	40
B	10	6	60
C	12	3	25
D	5	5	100
E	8	4	50
Totals	40	20	50

Jan. 1991 Source: Information from branches.

purposes of calculation or when comparing one list with another. In some cases assumptions have to be made, at other times there may be facts which can be used to decide on the limits. If the school leaving age is 16, then it can be assumed that the first class would be over 16 years and under 20 years. If the company has a strictly enforced policy of retirement at 65, the upper class would be 60 years but under 65.

These are all matters of judgement and it is useful to provide whatever footnotes are necessary to explain the basis on which a table has been constructed.

Interpreting statistical tables

Even the simplest table may contain a mass of information,

especially for a manager who knows the background and context of the table. In Table 2.2 (see the box), for example, Branch C has the most telephoned orders but has relatively few of these converted to confirmed orders, while Branch D has a 100 per cent record. This may be because of the location of the branches, or due to the follow-up procedure used by the different branches. It could be that there is a salesman in Branch D who is expert at converting prospective orders into firm ones. At the same time, the performance of Branch A could be investigated because of its low initial and confirmed orders. This table can provide the starting point for a detailed follow-up.

Managers are often presented with tables about which they have very little specialist knowledge and there is a need to

Table 2.3 *Output per person employed*

	Average annual % changes		
Whole economy	*1960–70*	*1970–80*	*1980–88*
UK	2.4	1.3	2.5
US	2.0	0.4	1.2
Japan	8.9	3.8	2.9
Germany	4.4	2.8	1.8
France	4.6	2.8	2.0
Italy	6.3	2.6	2.0
Canada	2.4	1.5	1.4
G7 average	3.5	1.7	1.8

UK data from Central Statistical Office. Other countries' data from OECD except 1988 which are calculated from national GNP or GDP figures and OECD employment estimates.

	Average annual % changes		
Manufacturing industry	*1960–70*	*1970–80*	*1980–88*
UK	3.0	1.6	5.2
US	3.5	3.0	4.0
Japan	8.8	5.3	3.1
Germany	4.1	2.9	2.2
France	5.4	3.2	3.1
Italy	5.4	3.0	3.5
Canada	3.4	3.0	3.6
G7 average	4.5	3.3	3.6

UK data from Central Statistical Office. Other countries' data from OECD, except France and Italy which use IMF employment data. 1988 data for France and Italy cover first three quarters only.
Sources: CSO, OECD, IMF

[*Treasury Economic Progress Report*, April 1989]

develop skills in gleaning all the information that is available in the table.

In Table 2.3 the output per person employed is shown for the Group of Seven industrial countries. Without knowing the background to the figures it is still possible to note that the output in the UK grew faster during the 1980s than in any other leading industrial country except Japan. The growth in output in manufacturing industries at 5.2 per cent is particularly marked and is over three times as great as the 1.6 per cent figure for the UK in the 1970s. The whole economy figure for the UK of 2.5 per cent for the 1980s is twice that of the 1970s at 1.3 per cent.

Assuming that the figures can be compared from the different sources (CSO, OECD and IMF) the growth of productivity was low in the UK in the 1960s compared with most of the other countries. The productivity growth in the other countries, except in the US, was lower in the 1980s than in the 1970s, while UK productivity was growing and the UK manufacturing industry productivity growth rates in the same periods were higher than in any other leading industrial country.

Management reports

The use of reports

Reports form an early stage of presentation, produced alongside or immediately after tables have been constructed; they form one of the primary methods used to interpret statistical data. There is a tendency to translate statistical data into a written form in order to make them more comprehensible and accessible. In fact, complicated data can very often be given sensibly only in numerical form; this includes many facts about sport such as league tables and batting averages and many facts about money, such as balance sheets and budgets. People interested in these facts learn to use the tables without difficulty and yet may feel more comfortable with other, unfamiliar, statistical facts translated into a verbal form.

In comparing the reports on the government annual budget in the newspapers it can be noted how little is reported in figures and how much in words, diagrams and cartoons. Textual reports are often the simplest method of presenting data and the easiest to understand for people not used to assimilating facts from statistical tables.

Reports can help to:

i interpret the information contained in tables and diagrams
ii emphasise the most significant points
iii explain the background to the information that has been collected
iv provide ideas about further action.

However factual they may attempt to be, reports involve immediate interpretation of the data. The order in which points are made is a matter of judgement, the vocabulary used to translate figures into words may indicate the relative significance of particular factors, the facts which are omitted may be as important in some circumstances as those included in the report. The manager writing a report may add points which are not in the numerical data but which are known to him from a different source. In reading a report the manager needs to keep in mind the expertise and background of the author.

Effective report writing

Effective report writing involves a clear presentation of the facts in a concise form which is as accurate and as brief as possible given the needs of communication. Technical language can be used where the report is for a specialist audience; for general purposes such language has to be avoided or explained. The same information may be reported in one way for in-company distribution and in another way for shareholders.

The usual advice is that reports should be set out in the following order:

i *Introduction*: aims and objectives of producing the report and terms of reference
ii *Main body*: contents of the report in a descriptive form with discussion and analysis
iii *Conclusions*: drawing together the most significant facts arising from the main body of the report
iv *Recommendations*: an action plan of what to do next
v *Appendix/Summary/References.*

If these are the 'rules', they may at times be broken to good effect. If the emphasis needs to be on action, then the recommendations could be put first, followed by the way they have been reached. Tables and diagrams may be included in the main body of the report or in an Appendix depending on their significance. One technique is to put the detailed table in the

Appendix with the main figures brought out in smaller tables in the main report.

Brevity is an important attribute for most company reports and a long report can start with an 'executive' summary, on the grounds that this at least will be read by most people and if the ideas catch the interest of the reader, the full report will then be read.

Frequency distributions and histograms

Frequency distributions

An early stage of presentation very often includes a frequency distribution and its presentation. Frequency distributions show how often a particular variable occurs, and many investigations and statistics are in this form. A manager may observe that customer complaints fall into four distinct categories and that it is possible to draw up a list as follows:

Categories of customer complaints	*Frequency of weekly complaints*
Advertising	15
Product	16
Service	34
After-sales service	5

A form may be designed so that the customer service department can simply note the number of complaints in each category:

Category		*Frequency of complaints*
Advertising	~~IIII~~ ~~IIII~~ ~~IIII~~	15
Product	~~IIII~~ ~~IIII~~ ~~IIII~~ I	16
Service	~~IIII~~ ~~IIII~~ ~~IIII~~ ~~IIII~~ ~~IIII~~ ~~IIII~~ IIII	34
After-sales service	~~IIII~~	5

This form can be used to construct a table with derived statistics (complaints about service represent 48.6 per cent of the total), and/or as a starting point for a report.

It is also possible to represent this frequency distribution by producing a diagram. The diagram which represents a frequency distribution is called a histogram. This name is derived from the Greek word *histos*, or mast, so that it is a mast diagram or bar diagram.

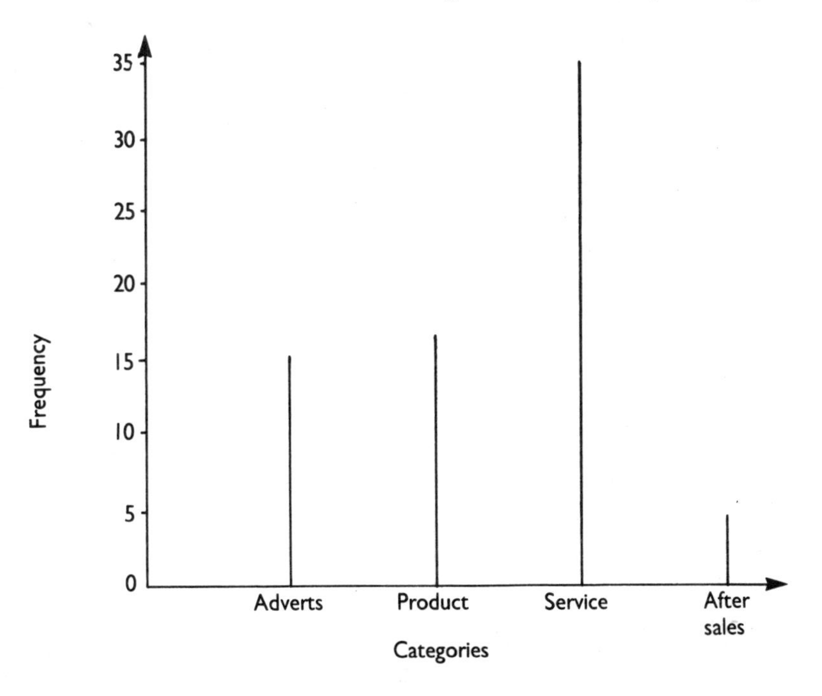

Figure 2.1 *Line chart*

Line charts

Figure 2.1 is a line chart (or mast chart) in which the length (or height) of each line represents the frequency. A histogram is an extension of the line chart and it is used when there is a grouped frequency distribution.

Histograms

These consist of a series of blocks or bars, each with an area proportional to the frequency.

Analysis of a company's wages and salaries bill may reveal the figures shown in Table 2.4. This can then be illustrated by a histogram (see Figure 2.2).

The areas of the blocks in a histogram are proportional to the frequency. The area is found by multiplying the width (the class interval) by the height (the frequency). In Figure 2.2 the area of the last block is $500 \times 5 = 2,500$.

Table 2.4 *Company pay analysis*

Monthly wages and salaries (£)	*Number of employees*
Up to 500	10
500 but less than 1,000	120
1,000 but less than 1,500	250
1,500 but less than 2,000	80
2,000 but less than 2,500	35
2,500 but less than 3,000	5
	500

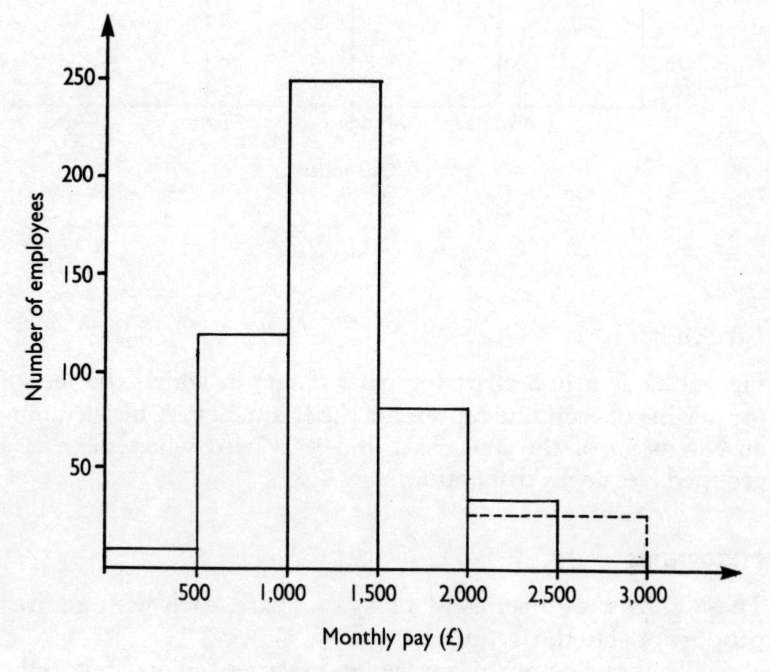

Figure 2.2 *Histogram*

The areas of the blocks must maintain their proportional relationship with the frequencies. This is particularly important if the frequency classes are uneven. If the last two classes in Table 2.4 had been amalgamated, the class interval would have been £2,000 but less than £3,000 and the frequency 40 employees. Separately, the area of the two blocks would be:

$$\begin{aligned} 500 \times 35 &= 17{,}500 \\ 500 \times 5 &= 2{,}500 \\ \hline &\ 20{,}000 \end{aligned}$$

and the histogram block would need to be drawn in proportion so that the width (class interval) was £1,000 and the height (frequency) was 20: 1,000 × 20 = 20,000.

Frequency curves

The next stage of representation is to draw a frequency polygon. This is a diagram which is drawn by joining up the mid-points of the tops of histogram bars. It is a 'many-sided figure' or polygon when the points are joined with straight lines and the horizontal

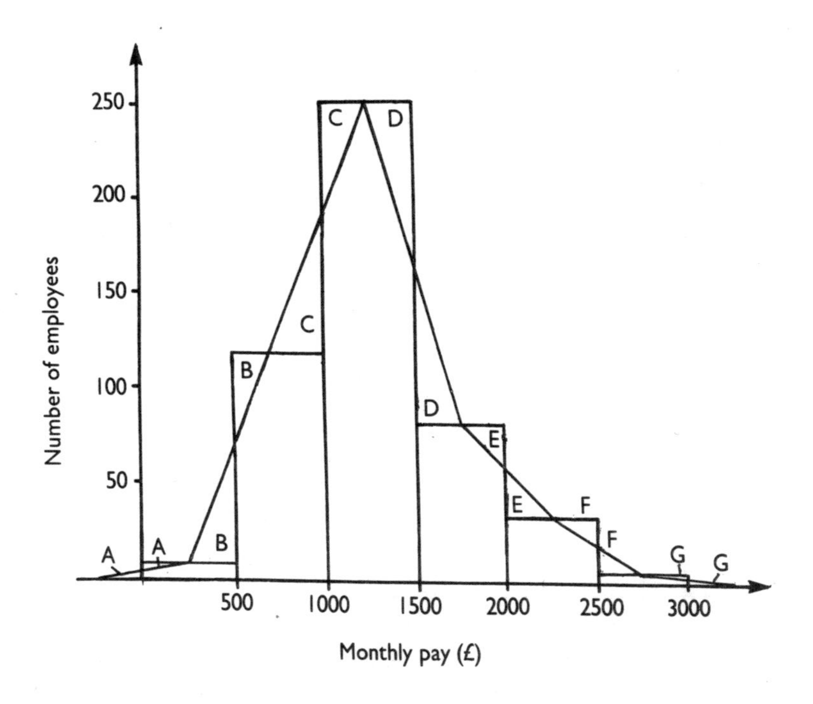

Figure 2.3 *Frequency polygon*

axis is included with the lines extended beyond the original range of the variable.

In fact the lines should be extended as they are in Figure 2.3 because the area of the polygon should be the same as the area of the histogram and this will only be the case if the triangles cut off the histogram by the lines of the polygon are compensated for by the triangles added by these lines (see the lettered triangles in Figure 2.3).

A frequency polygon can be drawn, without constructing the histogram first, by reference to the frequency and the mid-point of the class interval. The advantage of doing this is to have a general idea of the 'shape' of the distribution and to compare this with other distributions. Figure 2.4 shows two distributions based on similar data. It is obvious from these that the distribution of wages and salaries in Company A is 'skewed' or 'weighted' towards the lower end of the pay scales, while Company B is weighted towards the higher end. The shaded area highlights the

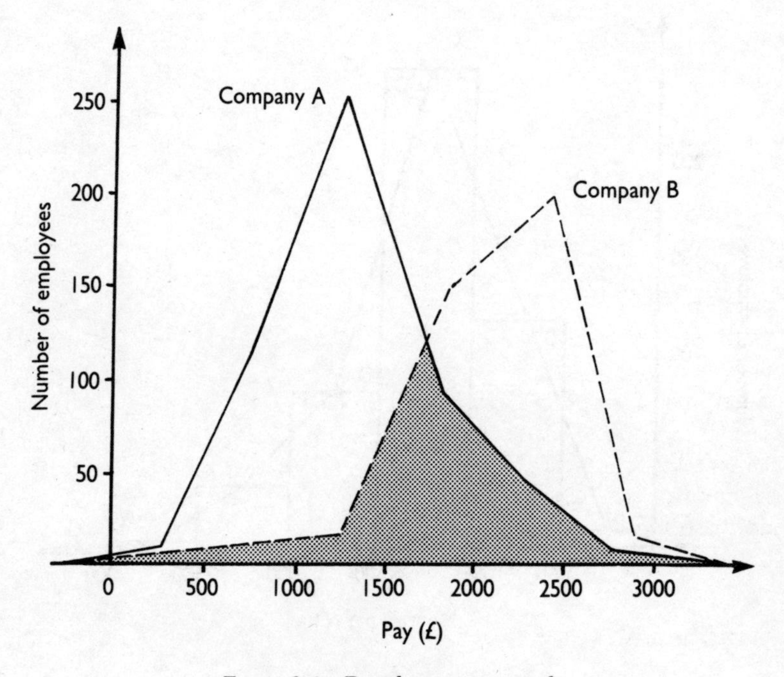

Figure 2.4 *Distributions compared*

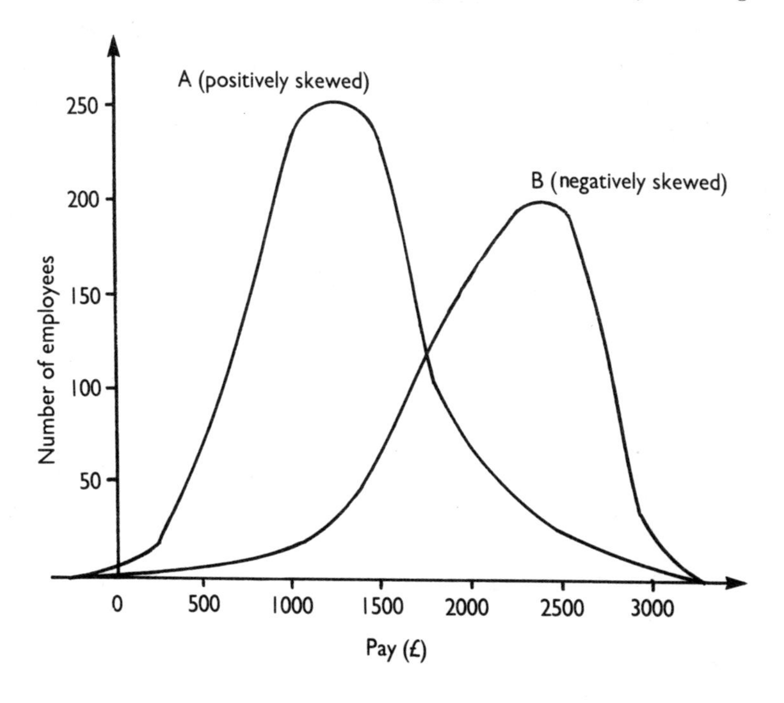

Figure 2.5 *Frequency curves*

pay scales the companies have in common and those that are different.

The straight lines may give an impression of a high level of accuracy while in fact the gradations between payments may be very much less sharp. A smooth curve may be a better representation of the original data or at least may not give a false impression of accuracy and it is for this reason that frequency curves are drawn. The representation of most distributions by diagrams or graphs is in the form of a curve for similar reasons.

The frequency curves in Figure 2.5 can be described as 'skewed' in the sense that their peak is displaced to the left or the right of centre. Distribution A is displaced to the left of centre and can be described as positively skewed and distribution B is displaced to the right and can be described as negatively skewed.

Graphs and their uses

These are normally drawn with smooth curves for the same

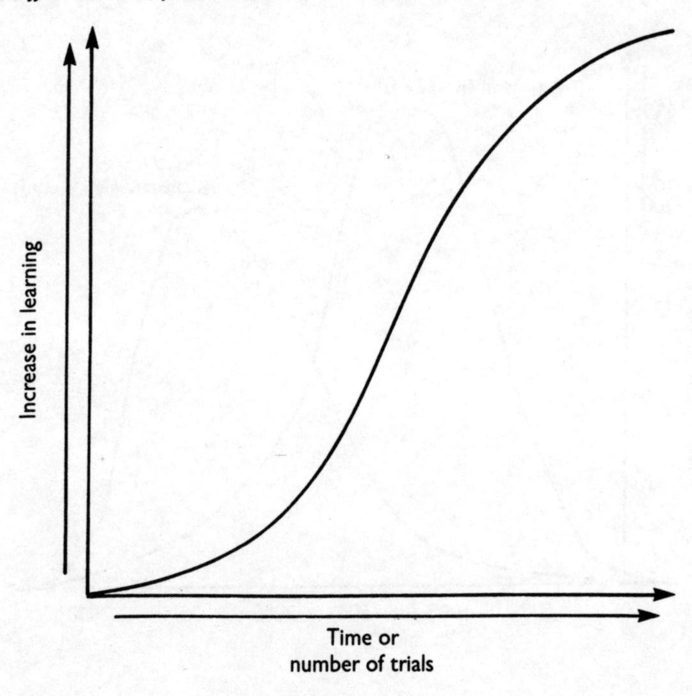

Figure 2.6 *Learning curve*

reason as frequency distributions. Graphs are drawn to illustrate the relationship between two variables and the word can be associated with 'graphic' or a 'vivid representation'. They can provide an immediate impression of a relationship.

Managers use graphs to show the relationship between such variables as demand, supply or price, sales and growth over a period of time. They are often used to show trends, as in Figures 2.6 and 2.7.

The learning curve in Figure 2.6 shows the link between learning a new skill and the time taken or number of trials or tests completed. The usual pattern is a slow start followed by a rapid increase in learning which levels out over a period of time. This is true of learning practical skills such as bricklaying, typing and drawing as well as learning about cerebral activities such as management and statistics.

The 'life cycle' of a product can be illustrated by a graph as in Figure 2.7 which shows the sales of the product (the dependent variable) over a period of years (the independent variable). The

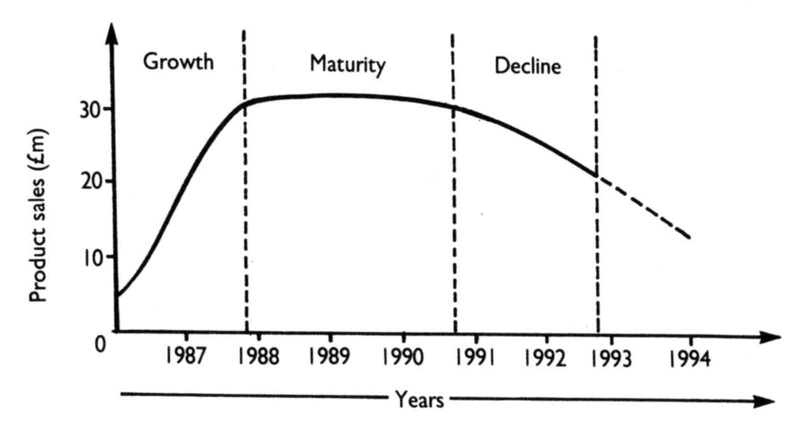

Figure 2.7 *Life cycle curve*

dependent variable is affected by changes in the independent variable; while the *independent variable* is not affected by changes in the dependent variable. Sales will change over the years, the years will not change in response to sales.

In practice it is not always so easy to decide which variable is dependent and which independent. The level of profit may depend on the level of investment just as investment decisions may depend on profits. It is a matter of judgement for a manager to decide which of the two is the more dependent.

The position of any point on the curve of a graph is decided by reference to the axes, and the points where the variables intersect are called 'bearings' or 'co-ordinates'. A set of points is built up and these are joined to form a curve. A point is fixed on a graph, like a map, by reading along the horizontal axis and drawing a vertical line up from this point and then reading up the vertical axis and drawing a horizontal line along from this point ('along the corridor and up the stairs', as in reading a map).

In Figure 2.7 it is possible to estimate that the level of sales in 1991 was at a level of approximately £31 million. This can be called 'interpolation' and when the direction of the curve is predicted into the future it is possible to 'extrapolate', for example, that in 1994 sales will be approximately £10 million. Any extrapolation needs to be made with great caution because the trends can change due to a range of factors which may be outside a manager's control. A competitor could close down, for example, and cease production, so that sales suddenly increase, or there could be a change in government policy creating a greater or lesser demand. If interest rates rise, so that credit is more expensive, the demand for the product may fall very rapidly. Any extrapolation of a graph will be based on assumptions. A manager may have to assume that interest rates will not change and that the competition will remain at the same level during the period of the extrapolation. Forecasts are based on facts and trends as they are known at the time, and on reasonable assumptions. However, with hindsight these assumptions may be proved to have been wrong.

The abuse of graphs

It is essential that graphs are drawn 'correctly', that is, without distortion. In 'persuasive' advertising, for example, graphs may be used to suggest something which is not strictly accurate. If the

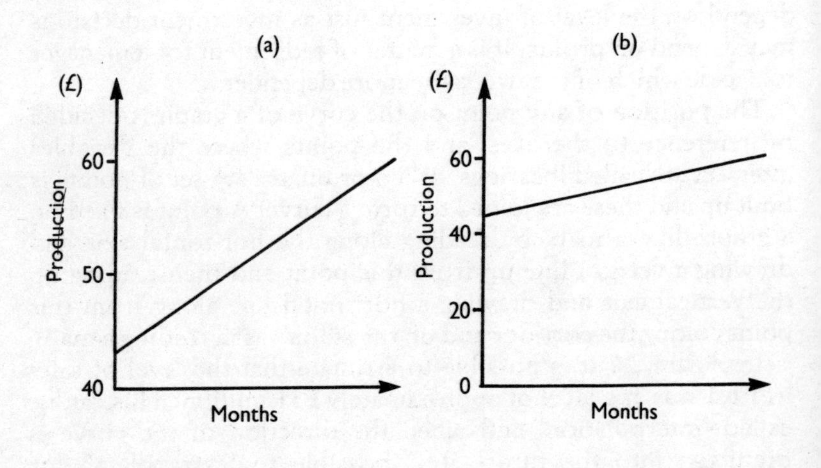

Figure 2.8 *Distorted graphs: point of origin*

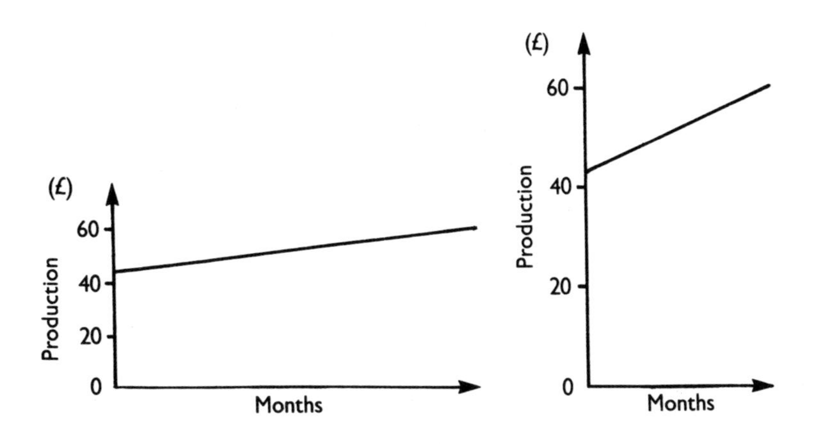

Figure 2.9 *Distorted graphs: axis*

vertical axis is not started at zero, the curve appears steeper than where zero, or the origin, is shown. In Figure 2.8(a) the growth in production appears much more rapid when the zero is not shown. The same effect can be achieved by compressing the vertical or horizontal axis. In Figure 2.9 the curves are flattened or sharpened by the compression of the vertical axis.

These forms of distortion are against the basic aims of statistical presentation, which are clarity of communication and accuracy in the presentation of information. This means that the vertical axis should always start at zero so that a false impression is not created. If it is not practical to have the whole scale running from zero then the scale can cover the relevant figures providing that the zero is shown at the bottom of the scale and a definite break in the scale is shown.

In Figure 2.10(a) the vertical scale is broken by a zig-zag to show that the line has been compressed. In Figure 2.10(b) the

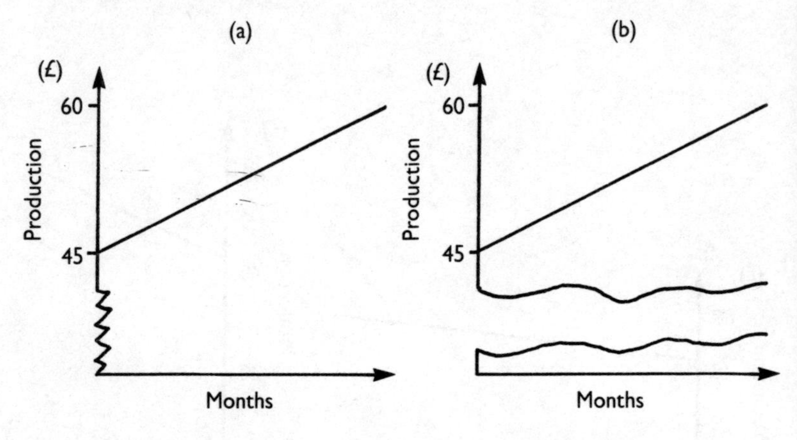

Figure 2.10 *Correct graphs*

same position is shown by the jagged lines running across the diagram.

Pictorial presentation

Bar charts and their uses

Bar charts are a commonly used form of presentation because they can illustrate simple information clearly and provide an immediate visual impact. Most information can be shown in the form of a bar chart of one type or another and like graphs they are easily produced through computer programs and desktop publishing.

Bar charts are more like a line chart (see Figure 2.1) than a histogram, because it is the height or length of each bar which is important. The height or length represents the data and the width and the area are not important because they are not drawn in proportion to the data. It is for this reason that the bars on any particular chart are usually drawn the same width. They can be drawn vertically or horizontally, as in Figure 2.11(a) and (b),

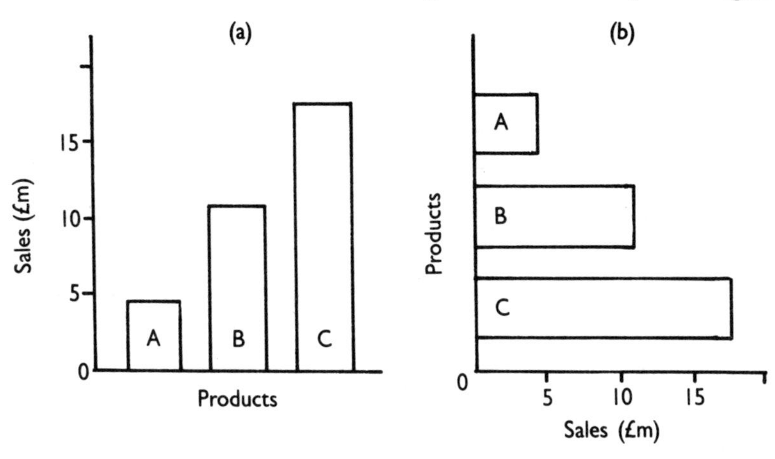

Figure 2.11 *Simple bar charts*

depending on which is felt to make the best visual impact. They can be drawn 'independently' or put on to a scale, although unlike a graph, interpolation and extrapolation is not really possible. Simple trends over a period of time can be seen at a glance, while the main comparison is between each 'discrete' bar.

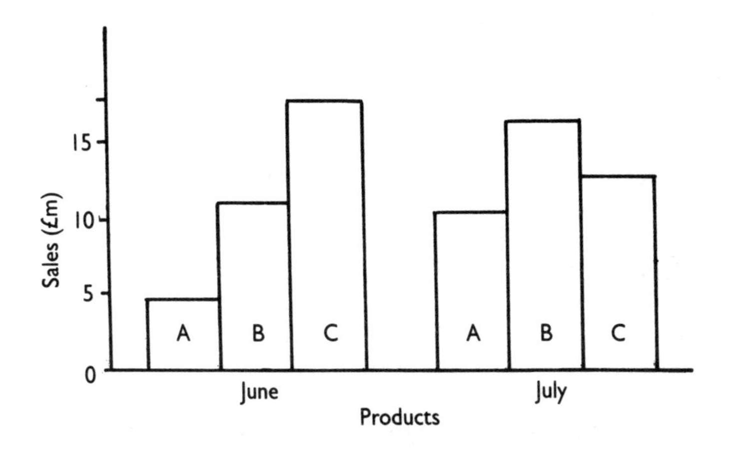

Figure 2.12 *Compound bar chart*

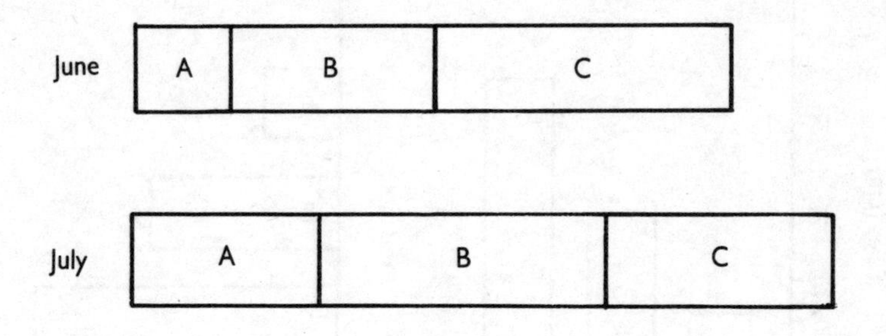

Figure 2.13 *Component bar chart*

There are a number of variations on the simple bar chart which can help to illustrate and to emphasise particular characteristics of the items.

Variations on the simple bar chart

Compound bar charts: Compound or multiple bar charts can be used to compare a number of items within, say, a month as well as comparing the items between the months.
Component bar charts: Component bar charts are useful in showing the division of an item into its constituent parts. Figure 2.13 shows the same information as Figure 2.12 but it accentuates the proportions of total sales rather than the growth and decline of the items over time.
Percentage component bar charts: Percentage component bar charts emphasise the proportional changes between items to an even greater extent, while not showing the changes in total sales. Each bar is the same length to show 100 per cent of the sales for that month.

Pie charts

These are circular charts with the circle representing the total quantity of a variable: total sales, or output or costs. The circle is usually divided into sections to represent the parts of the whole.

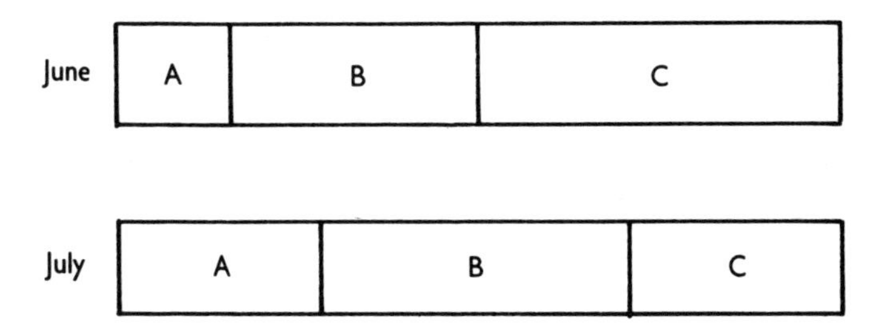

Figure 2.14 *Percentage component bar chart*

They are called pie charts because the area of a circle is πr^2, π being the Greek pi; alternatively it has been suggested that they look liked a sliced pie—as in apple.

In theory, each sector of the circle should have an area equal to the quantity of the variable and if two pie charts are used for comparison, the areas should be proportional. In practice, many pie charts used by organisations are produced for illustration only and the proportions are not necessarily accurate. They are useful when a few items make up proportions of a whole and where the proportions are more important than the numerical values.

A manager may want to illustrate the proportion staffing costs make up of the total, as in Figure 2.15.

If it is assumed that the proportions in Figure 2.15 have been calculated, then they would be as shown in Table 2.5.

There are 360° in the area of a circle and therefore capital will occupy:

$\frac{10}{100} \times 360 =$	36°
Power etc. will occupy	36°
Materials will occupy	72°
Staff costs will occupy	216°
	360°

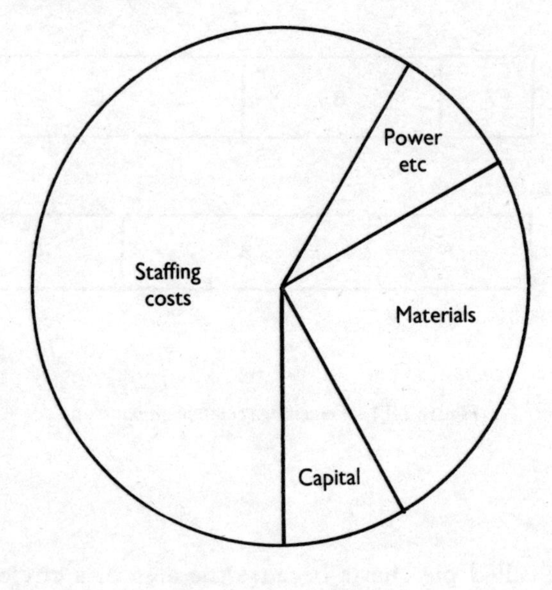

Figure 2.15 *Pie chart*

Table 2.5 *Quantified staffing costs in relation to total costs*

Costs	*£m*
Capital	10
Power, maintenance, transport	10
Materials	20
Staffing	60
Total	100

Even without the detailed information, it is possible to see at a glance from Figure 2.15 that staff costs are more than half of the total, power and capital each make up similar proportions of the

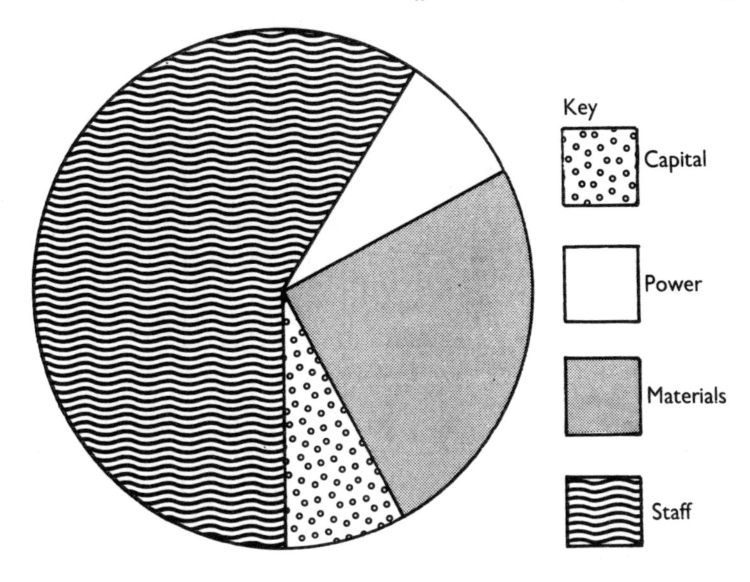

Figure 2.16 *Another pie chart*

total and materials costs are in between. If detailed information is required a table will easily supply it.

Figure 2.16 is an illustration of another method of presenting a pie chart.

Pictograms and their uses

In one sense all diagrams are pictograms because they are pictorial illustrations of data. The term is applied more directly to *picto*rial dia*grams*. They are pictures used to represent data. They should provide an immediate visual impression and therefore need to be kept very simple.

Managers can use pictograms to illustrate reports, so that the growth in output from one year to another can be shown by units of the goods produced (as in Figure 2.17). To add a little more information, the number of employees needed to produce

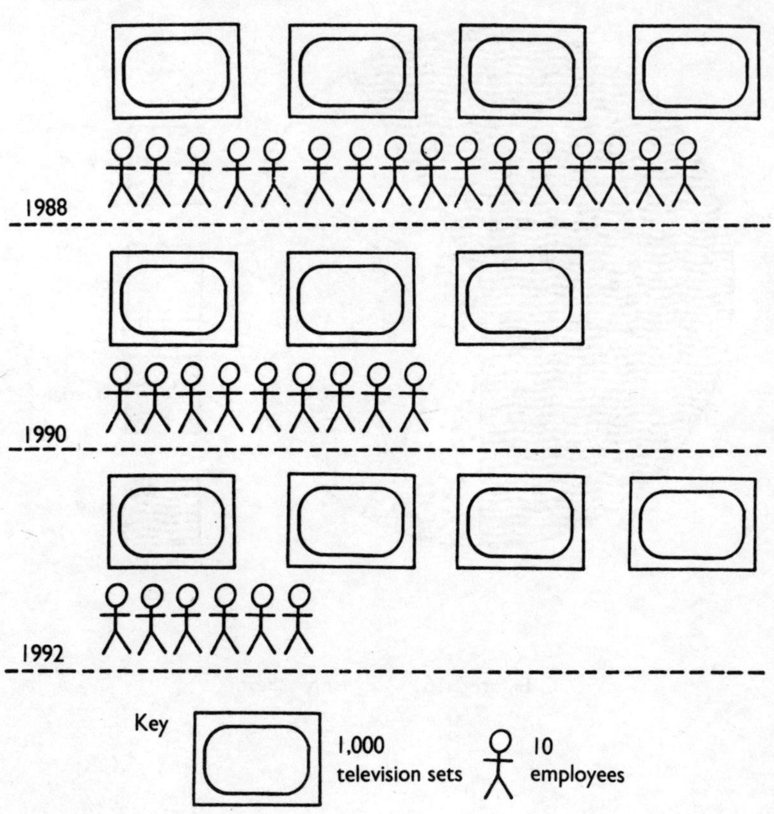

Figure 2.17 *Pictogram*

this output can be shown in order to provide some impression of productivity.

Pictograms can include symbols (such as squares) to illustrate items and can be used in conjunction with any other form of pictorial presentation. Maps are often used, for example, to illustrate trade reports or sales figures for regions of countries. Graphs, symbols, pictograms, flags and so on can be superimposed on these 'map charts' or 'cartograms' to represent various items.

Company reports, advertising and public relations literature, brochures and prospectuses often contain examples of these forms of presentation in a variety of combinations. They should be used where they are the best form of communication.

Management diagrams

These are diagrams particularly used in the management of any organisation. They can be statistical in the sense that they show numerical relationship (product life cycles, for example) or descriptive in showing relationships such as lines of communication. Management or organisation charts, for example, can illustrate relationships between staff. The traditional 'pyramid' chart shows 'line' relationships in a hierarchical form, so that unbroken lines from the top to the bottom of the organisation show how authority is exercised through each level from the managing director or the board of directors down to the newest junior clerk or apprentice (Figure 2.18).

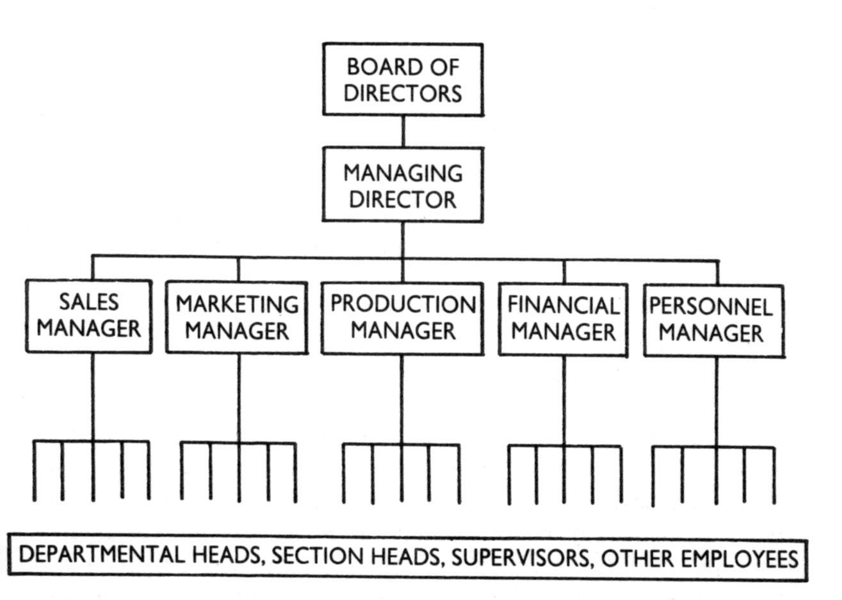

Figure 2.18 *Management chart*

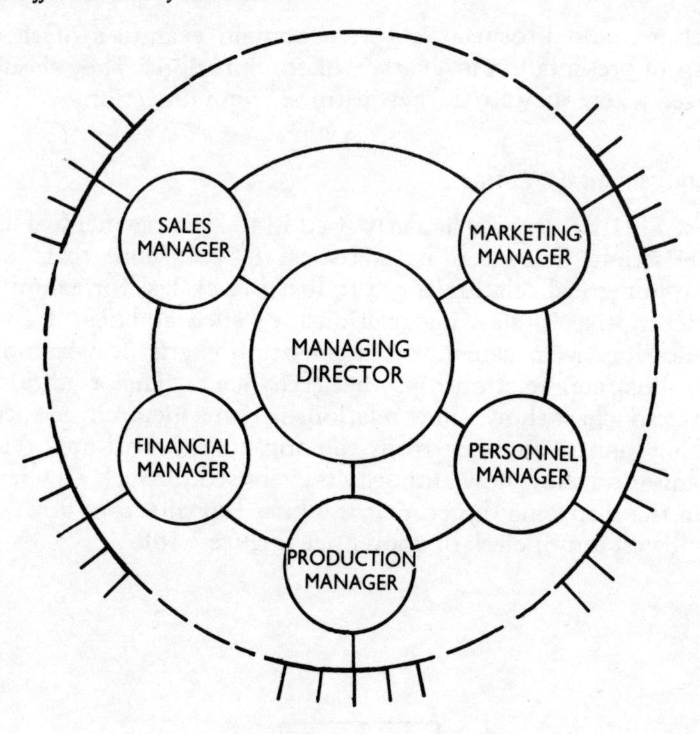

Figure 2.19 *Circle diagram*

This chart suggests clear lines of authority so that the subordinates know who is their boss and the bosses know who is under their charge and each manager is shown to be responsible for an area of work within the organisation. Another way of showing this relationship is a circle diagram as in Figure 2.19.

Where managers have functions across the organisation there may be an overlap of responsibility and subordinates may have more than one boss. At its extreme this is a matrix management structure (Figure 2.20), with some line managers concerned with traditional 'vertical' functions while other line managers are concerned with aspects of these on a 'horizontal' plane. The personnel manager, for example, may be responsible for all staff training, all staff personnel matters, all industrial relations and the recruitment and selection of staff. The production manager will then have responsibility for the technical aspects of production only, and some staff matters such as team building and

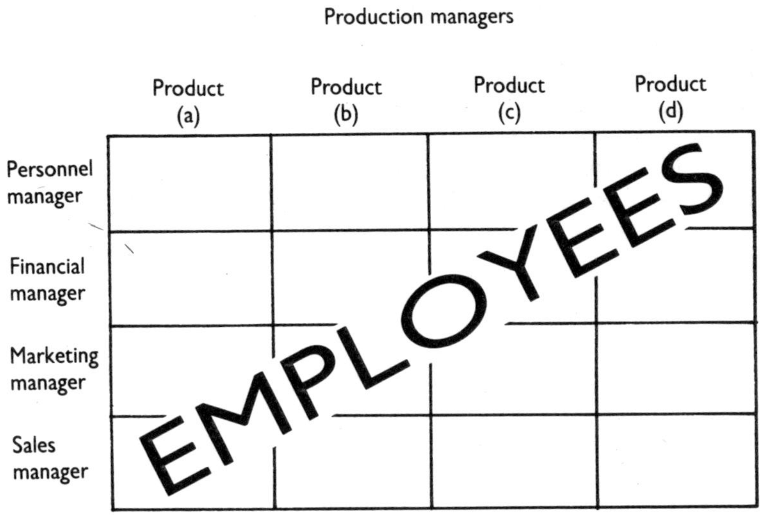

Production managers

	Product (a)	Product (b)	Product (c)	Product (d)
Personnel manager				
Financial manager				
Marketing manager				
Sales manager				

EMPLOYEES

Figure 2.20 *Matrix structure*

promotion, while other staffing matters and all marketing, design, purchasing and so on are carried out by other managers in the matrix organisation.

- The purpose of *presentation* is to inform, analyse and persuade.
- *Effective presentation* is the skill of using the best format for a particular audience at a particular time.
- *Forms of presentation* include line charts, histograms, frequency curves, graphs, bar charts, pie charts and pictograms.
- *Statistical tables* are constructed from raw data and involve classification and interpretation.
- *Qualitative data* depend on judgement and subjective criteria.
- *Quantitative data* include variables with measurable characteristics.

- *Discrete variables* are measured in single units.
  Continuous variables are in units of measurement which can be broken down into definite gradations.
- *Management reports* are used for the presentation and interpretation of statistical data.
- *Frequency distributions* show how often a particular variable occurs.
- A *histogram* represents a frequency distribution.
- *Graphs* are drawn to illustrate the relationship between two variables.

3
The Art of Collecting Management Information

Management information

If data are not already available and they are needed to help to solve a problem, they have to be collected, and this involves carrying out a survey or inquiry of one type or another. Management involves a constant flow of primary information as a natural part of the job. Directly a manager asks 'why?' the answer will be in the form of primary data. 'Why was production low yesterday?' 'Because of staff absences, one of the machines was not working, some raw materials did not arrive.' These explanations are all the starting point for an inquiry.

Management surveys

A very limited inquiry can be carried out very quickly by observation. It may be possible, for example, to carry out a survey on staff absenteeism in a factory by a quick observation of the machine shop. This may show that out of 50 machines, 8 are idle and do not have a machine operator present. It could be stated, therefore, that the absenteeism rate is 16 per cent.

This is a limited survey given a certain spurious validity by using a percentage. Detailed investigation could prove that 5 out of the 8 missing operatives were on a scheduled tea break, 1 was in the maintenance store collecting a spare part and 2 were in fact absent. The actual absenteeism rate was 4 per cent and the loss of production was 6 per cent, because scheduled meal breaks were considered part of the total production costs, while 2 per cent of the loss of production was due to a faulty machine.

The original survey was so limited in both time and frequency that it could not be used to provide a general picture of the absenteeism rate in the factory, although it can be argued that some type of survey, however limited, is better than none at all. It is a starting point. The more detailed investigation inevitably presents a more complex picture and if this detail were combined with regular observations, results could be produced which would be useful to the management of the factory.

Argument over the level of absenteeism could be narrowed, a sense of proportion established and the statistical evidence could help in making a decision about what to do about absenteeism.

Stages in survey design

Surveys

The longer and the more systematic a survey, the more chance there is of its being useful for management. Whatever the size and the level of detail in a survey, there are a series of stages in carrying it out:

i *the survey design*: this will depend on the subject of the survey, the amount of time and money that can be spent on it, and the accuracy required. The observation of the machine shop on one occasion is quick and cheap but not very accurate. The UK Census of Population, on the other hand, aims to include everybody in the country and to find out a range of information from every household. It is time consuming and expensive, but can produce a large amount of fairly accurate information. It is a 100 per cent survey, while the observation was a sample.

ii *the pilot survey*: this is the next stage in carrying out a large survey. It is a preliminary survey on a small scale to make sure that the design and methodology of the main survey are likely to produce the information required. It is a way of testing the survey design so that it can be refined and altered if necessary, before time and money have been spent on the main survey.

iii *the collection of the information*: the main survey methods are through observation, interview and questionnaire or a combination of these methods. These methods form part of the survey design and they are considered in detail below.

iv *classification and coding*: this is part of survey design to ensure that answers to questions can easily be tabulated. Questions are often pre-coded by giving them a reference number or letter.

v *presentation*: the final stage of a survey is usually to construct a table and write a report. It is useful to consider the problems that may arise in producing these at the time the survey is being designed.

Observation and its uses in management

Everybody uses observation in order to collect information. Most observation is unsystematic and informal and it is not necessarily used for any particular purpose in everyday life. It begins to become systematic when it is used for a purpose. Consumers will look into a number of shops, comparing prices, quality and style, before deciding what to buy. Managers may observe the level of absenteeism in their labour force in a fairly casual way before realising the importance of being systematic.

Direct systematic observation is a powerful method of collecting information. It can be an unobtrusive and objective method, by watching people's behaviour without their knowledge and recording the results. Road surveys are of this nature, where a particular section of road is observed and the number and type of vehicles using it are counted at various times of the day on different days. Systematic observation can be used to inquire into working habits in offices and factories in time and motion studies and to observe the movement of shoppers in supermarkets. Mechanical means of observation can be applied, with an electrical monitor, a television monitor on the roads, a closed circuit TV monitor in the supermarket. The caricature of the time and motion investigator with a stop-watch, clip-board, binoculars and a worried expression has given way to more sophisticated means of observation.

Problems in collecting information by observation

The main problems to the manager of using observation as a method of collecting information are:

i *objectivity*—the observer cannot easily remain objective and also ask the detailed questions which will help him or her to understand events

ii *selectivity*—observers can be unintentionally selective in what is observed and recorded, so that, for example, observing the emergency ward of a hospital does not necessarily provide a typical view of the work in the rest of the hospital. Observers may record only certain aspects of behaviour, those they consider to be important. On the other hand a total recording of events by television or radio may provide more information than it is possible to analyse iii *interpretation*—observers may decide what is happening in front of them without knowing the real reasons for an activity. This may be an example of the old problem of interpretation of the observation of a worker leaning on a spade, who is thought to be slacking by one observer, and taking a well-earned rest by another

iv *chance*—unless observations are taken frequently over a period of time, a chance event may be mistaken for a recurrent one. The flow of traffic on a stretch of road may be heavy on a particular day because there are roadworks blocking an alternative route

v *participation*—observers can influence people's behaviour when they realise they are under observation.

In spite of these problems, direct systematic observation is the classic method of scientific inquiry used by biologists, physicists and other natural scientists. In the social sciences it can be used as a method of watching humans objectively and at a distance, as if observing animals. This is, however, to ignore the great advantage that doctors have over vets, that they can ask the patient what is wrong with them and expect a helpful reply.

Interviewing and its uses in management

This method of collecting information makes full use of all means of communication and the gift of language. An interview can be described as a conversation with a purpose. In an informal sense everybody uses interviewing to obtain information. In the course of a general conversation a manager may ask a supervisor: 'Why hasn't this report been written yet?' to receive a list of excuses about time, other priorities and obstacles in the way of progress on it. The general question may be followed up with more detailed ones, such as: 'What other priorities?', 'What happened in the time allotted to you in order to write this report?' and so on. This may lead to further explanations and questions.

A similar approach is used in formal interviewing. A 'formal' interview is a conversation between two people that is initiated by the interviewer in order to obtain information. Although the original question by the manager to the supervisor was a casual one, as the questions continue the conversation becomes more like a formal interview. The interview will be more structured than the usual conversation, because the interviewer will present each topic by means of specific questions and will decide when the objectives of the interview have been satisfied.

Interviews are one of the most important ways in which managers obtain information. The most obvious example is the job selection interview. The usual process will be for an advertisement to have been placed, inviting applications. These are then reviewed, in order to decide on a long list or a short list to interview. The interviews are usually structured in such a way that similar questions are asked to all candidates with supplementary questions asked as necessary. The interviewer(s) will determine when it has been decided who will be offered the job and the objective of the interviews has been satisfied. It will not be the only criterion for selection. The application form and curriculum vitae will play an important part and managers may use personality tests, team problem solving or other forms of selection procedure.

Interviewing skills

There are skills in interviewing which can be learned. Managers will acquire these skills by training, usually including role-play techniques, by careful preparation and by experience. Role-play techniques are used in training because interviewing is an interactive method of finding information.

Interviewers differ in their skills, while respondents differ in their motivation and knowledge and the content of the interview differs in complexity. Interviewing is a subjective process and it depends on the people involved being able to communicate well. If they do not like each other or they have conflicting 'body language', the interview can fail to achieve its objective. At one extreme, interviews have been carried out with screens erected between the interviewer and the respondent in order to reduce the effects arising from 'non-verbal' communication. In many interviews there is a need to see the other person because it can be argued that the attitude of people, as shown by the way they sit or how they are dressed, is an important element of communication.

Interview questions

The way questions are asked, as well as who asks the questions, may influence the answers. An employee may answer similar questions differently if they are asked by a senior manager or by a colleague. Questions can be asked in such a way that they encourage agreement. This appears to be the explanation for canvassing returns which may show two parties with clear majorities. Voters may want to give answers most likely to please the canvasser.

Interviewer bias

Interviewer bias can arise from a subtle influencing of the results by:

i the way questions are asked
ii the extent of supplementary questions
iii the sequence of questions
iv the expectations of interviewers related to the age and appearance of the respondent.

Interviewers can make careless mistakes in the recording of answers or in the reporting of the results. Interviewers can fabricate answers in order to avoid the time and effort involved in collecting them.

The respondent

The respondent can also make a success or failure of the interview. This will depend on:

i *The level of motivation*: a market research interview carried out in the entrance to a supermarket may suffer from the fact that the respondent's main objective is to complete his or her shopping as soon as possible and not to answer a list of questions. An applicant for a job is much more likely to be highly motivated at interview.

The interviewer may be able to suggest why it is in the respondent's interests to answer questions. People will often answer questions at length if they believe that they can influence events, such as improving the surroundings of the firm's canteen or deciding on the site of a new supermarket.

Payment may provide an incentive, but it may be the wrong one. People may welcome answering questions simply for the money and perhaps without any knowledge of the subject.

ii *The respondent's role*: the role the respondent has in the interview needs to be clear in his/her mind and in the interviewer's mind. Respondents need to know if they are being asked questions as employees, experts, consumers, colleagues or potential job applicants, otherwise they will not know what information is relevant and how detailed the answers should be.

iii *The accessibility of information*: the information being sought by the interviewer has to be known to the respondent otherwise the answers are likely to be irrelevant.

Questionnaires and their uses in management

Questionnaires are lists of questions aimed at discovering particular information. They are a relatively cheap and easy method of collecting information and they are widely used, both by managers to obtain information for themselves and by other people in order to obtain information from managers. They can be used in interviews in order to standardise the questions and they can be used in observation in order to make it more systematic, by requiring the observers to answer a list of questions about what they are observing.

Questionnaires are frequently distributed by post. This enables distribution to large numbers of people to be carried out cheaply. At the same time the answers can be carefully considered. The main disadvantage of questionnaires sent by post is that there is usually a poor response. Again, it is a question of motivation and respondents need a strong incentive to return a completed questionnaire. Tax returns and the population census have legal penalties attached to them if they are not returned. Promotional questionnaires such as those produced by magazines and travel companies often attach prizes to the return of their questionnaire forms. Managers may be able to obtain a reasonable rate of response to questionnaires on matters of interest to them by suggesting that the survey is part of a consultation process and that, therefore, the manager may subsequently influence policy.

In order to overcome this problem of a poor response rate, questionnaires can be delivered by hand and then collected when completed. This puts some pressure on people to complete and return the forms, but it is a more expensive process than a postal survey and there will be a more limited distribution. Within an organisation, managers can insist on the filling in of questionnaires and these can form a useful aspect of consultation in the development of policy.

Question design

In all questionnaires, the design of the questions is a very important consideration. These should reflect the aims and objectives of the survey, so that there needs to be certainty about why it is being carried out, how the results will be used and what type of questions need to be asked in order to achieve these objectives.

Questions should be:

- i simply and clearly worded so that all possible respondents will understand them
- ii useful and relevant in order to produce the desired information
- iii free from bias
- iv in a logical order, with similar subjects grouped together and one question leading on to another

Table 3.1 *Population census (a)*

H3 Amenities

Has your household the use of the following amenities on these premises?
Please tick the appropriate boxes

- A fixed bath or shower permanently connected to a water supply and a waste pipe

1 □ YES—for use only by this household
2 □ YES—for use also by another household
3 □ NO fixed bath or shower

- A flush toilet (WC) with entrance inside the building

1 □ YES—for use only by this household
2 □ YES—for use also by another household
3 □ NO inside flush toilet (WC)

- A flush toilet (WC) with entrance outside the building

1 □ YES—for use only by this household
2 □ YES—for use also by another household
3 □ NO outside flush toilet (WC)

[Office of Population Censuses and Surveys]

v unambiguous and free from 'vague' words such as 'probably', 'generally'
vi capable of being answered easily and quickly, preferably with a tick/cross or by putting a letter or number in a box.

Questions should not be:

i too personal otherwise respondents may be reluctant to answer
ii leading (for example, 'don't you think this should be stopped ...?' encourages a positive answer)

Table 3.1 *Population census (b)*

Whether working, retired, housewife etc. last week

Please tick all boxes appropriate to the person's activity last week.

A *job* (box 1 and box 2) means any type of work for pay or profit but not unpaid work. It includes:
casual or temporary work
work on a person's own account
work in a family business
part-time work even if only for a few hours

A *part-time* job (box 2) is a job in which the hours worked, excluding any overtime, are usually 30 hours or less per week.

Tick box 1 or box 2, as appropriate, if the person had a job but was not at work for all or part of the week because he or she was:
on holiday
temporarily laid off
on strike
sick

For a full-time student tick box 9 as well as any other appropriate boxes.

Do not count as a full-time student a person in a paid occupation in which training is also given, such as a student nurse, an apprentice or a management trainee.

1 ☐ In a full-time job at any time last week
2 ☐ In a part-time job at any time last week
3 ☐ Waiting to take up a job already accepted
4 ☐ Seeking work
5 ☐ Prevented by temporary sickness from seeking work
6 ☐ Permanently sick or disabled
7 ☐ Housewife
8 ☐ Wholly retired from employment
9 ☐ At school or a full-time student at an educational establishment not provided by an employer
0 ☐ Other, please specify

....

[OPCS]

Once the questionnaires are collected they have to be processed into a form that can be useful as a basis for decisions. Tables will be produced from each question (such as: Yes 100, No 50, Don't Know 10, Unanswered 10), followed by a report and graphs and diagrams. A series of calculations may be made, such as averages, percentages, measures of dispersion and correlation, and perhaps trends can be observed, leading to forecasts and the formulation of policy.

The 1981 Population Census in England included questions about household amenities, type of work and daily journey to work. These are good examples of questions designed to be answered by a wide range of people and in a form that is easily processed:

Table 3.1 *Population census (c)*

Daily journey to work

Please tick the appropriate box to show how the longest part, by distance, of the person's daily journey to work is normally made.

For a person using different means of transport on different days show the means most often used.

Car or van includes three-wheeled cars and motor caravans.

1 ☐ British Rail train
2 ☐ Underground, tube, metro, etc
3 ☐ Bus, minibus or coach (public or private)
4 ☐ Motor cycle, scooter, moped
5 ☐ Car or van—pool, sharing driving
6 ☐ Car or van—driver
7 ☐ Car or van—passenger
8 ☐ Pedal cycle
9 ☐ On foot
0 ☐ Other (please specify)

....

0 ☐ Works mainly at home

[OPCS]

- *Management surveys* are designed to collect primary data: surveys are → to discover information → in order to reduce the area of argument → to help make decisions.
- *Survey methods* include observation, interviews, questionnaires and a combination of these methods.
- *Survey design* includes the choice of a survey method, a pilot survey, classification, coding and presentation of the results.
- *Observation* can be informal, systematic, mechanical and participant.
- *Interviews* are conversations with a purpose.
- *Questionnaires* are lists of questions aimed at discovering particular information.
- *Attention* needs to be given to:
  survey design, question construction, cost and time involved, aims and objectives.

4
The Science of Collecting Management Information

Probability and sampling

A survey can be carried out on a 100 per cent basis, including everybody or every item in a group of people or things about which information is required. Such a group is called a 'population'. A *sample, on the other hand, is anything less than a full survey of a population*. A 'population' could be, for example, all voters in the country for an opinion poll, or all employees in an organisation, or all items coming off a production line. A population can include, therefore, millions of people or items or a much smaller number.

A sample is usually thought of as a small proportion of the population which is surveyed in order to provide an idea of the quality of the whole. Sampling is based on the theory of probability and can only be understood in these terms. It may appear to be desirable always to base decisions on complete surveys including all the people or items involved; anything less than this might be felt to include only a part of the information and to be open to a high degree of error. In practice it is often possible to obtain more accurate results by carrying out a sample rather than a complete survey; a 'clean' (ie well designed, organised and processed) sample may be more accurate than a 'dirty' (ie poorly organised) survey.

The problem with a full survey is the cost and time involved in carrying it out. This will not matter in a small organisation or when only a few items are in the population. With a large population it is usually better to concentrate the resources available on investigating a sample of people or items rather than spreading resources thinly over all items.

The possibility of reaching valid conclusions concerning a population from a sample is based on two mathematical laws which are ways of describing the theory of probability. This is the basis of statistical induction, the process of drawing general conclusions from a study of representative cases.

i *The law of statistical regularity* states that a reasonably large sample selected at random from a large population will be, on average, representative of the characteristics of the population.

The sample must be collected at random, which means that every item in the population has an equal chance of being included in the sample, and the number of items in the sample must be large enough to represent the various aspects of the whole population. Opinion polls, for example, should include enough people to reflect the views of the various political parties. If ten people were asked at random which party they supported they might, by chance, all favour the same party. This would not represent the whole electorate if it was known that some people favoured one or other of the alternative parties. A poll of 100 people taken at random would be less likely to show unanimous support for one party and a sample of 1,000 might start to provide a good impression of the views of the population. While a sample of 1,000 or 2,000 might provide fairly accurate information, increasing this to 10,000 or 12,000 might not achieve a very significant improvement in accuracy but would greatly increase the costs of the sample survey.

ii *The law of the inertia of large numbers* states that large groups of data show a higher degree of stability than small ones.

This is because there is a greater tendency in a larger number of observations for variations in the data to be cancelled out by each other.

Sample size

A large sample will be more reliable than a small sample taken from the same population, providing that other things, such as design and processing, are equal. A large sample will be, however, much more expensive and greater and greater degrees of accuracy will need proportionally larger and larger samples. While the size of a sample does not depend on the size of the population, it does depend on the resources available or the degree of accuracy required. A population which is known to have great variations within it (people of different opinions or units of various types) will require a larger sample to represent it and to ensure that these variations are included, than a population known to be more homogeneous. In recent years the size of political opinion polls has tended to be increased because of the greater volatility of the electorate.

Sampling and non-sampling errors

These do not depend only on the sample size, they depend also on the sample design and on the nature of sampling itself.

Sampling error: this arises because a sample cannot exactly represent the population from which it is chosen. The degree of sampling error will depend on the size of the sample, the larger the sample, the smaller the sampling error. It is important to note that this does not depend on the size of the population. A population of 500,000 does not require a larger sample than a population of 250,000 or 25,000.

Sampling error is the difference between the estimate of a value obtained from a sample and the actual value. A sample may show that the average length of a component is 20 cm when the actual average is 20.5 cm. The sampling error is 0.5 cm. This is only known, of course, when a full survey is taken and this does not usually happen. At election times comparisons are made between the pre-election opinion polls and the actual result. Apart from sampling error there is a time difference and other factors which may influence the differences in the results. As in other aspects of statistics, the interpretation of sample results is a matter of judgement.

Non-sampling error: this arises due to problems with the sample design. This kind of error includes the choice of sampling frame (see p. 69) and sampling units and the degree of attention given to detail in the survey.

The total error arises from both non-sampling and sampling errors and they both need to be controlled simultaneously. A larger sample will not reduce the total error if there are design faults in the sample.

Sample design and its objectives

The fundamental principle of sample design is to avoid bias in the selection procedure. This is achieved by random selection of the people or units in the sample. If a 'sample' is selected by other means the law of probability will not apply and the results achieved will not be statistically valid.

The other major factor in sample design is a practical one, in that the objective of sample design is to achieve the maximum precision for a given outlay of money and time.

Designing a sample

i It is essential to decide on the objectives of the survey and what it is aimed at achieving.

ii It is essential to define the sample population and the sample unit, that is the group of people, items or units under investigation.

iii The people or units have to be defined in terms of particular characteristics. A sample survey within a company may be taken on part-time employees only and they need to be clearly identified. Does part-time mean anything less than full-time? If this is the case, the population may include people working nearly full-time and others working a few hours a week.

iv The sample frame has to be selected. This is the list of people, items or units from which the sample is taken and this is chosen once the sample unit has been defined. This could be, for example, the wage list for all part-timers in a company. The frame should be comprehensive and up-to-date. General examples, used in national surveys, are the electoral registers, telephone directories, magazine subscription lists and wage lists.

v The survey method has to be decided just as in a full survey. Sample surveys can include observation, interviews and questionnaires and combinations of these methods.

Bias in sampling design

Bias consists of non-sampling errors, which cannot be reduced

Table 4.1 *Accuracy in information collection*

Problems can arise because of:
Poor question design
Interviewer bias
Lack of respondent motivation
A sample which is too small
Poor sample design
The use of a faulty sampling frame
Poor processing
The misinterpretation of results
The 'human factor'

or eliminated by an increase in sample size. It arises from an inadequate sampling frame, a sample not selected on a random basis, poor or ambiguous question wording, and a 'non-response' from particular groups. Some sections of the population may be hard to find or may refuse to co-operate and will be under-represented in a survey. It may be, for example, that a survey on part-time workers includes few of those working only two or three hours a week, because they are difficult to contact at work. The problems of accuracy in information collection are summarised in Table 4.1.

Sampling methods

These fall into two categories, those which are fully random (simple, systematic, stratified) and those which are non-random (multi-stage, quota, cluster). In using non-random methods there is usually an attempt to include some element of randomness because it is difficult to assess the sampling errors involved in non-random samples.

Simple random sampling

Each unit (or person) in the population has the same chance as any other unit of being included in a simple random sample.

A lottery method is used for selection, for large populations usually on a computerised system of random numbers, as used for British Premium Bonds and state lotteries. A 10 per cent sample of an organisation's 8,000 employees would be carried out by giving each employee a number from 1 to 8,000 and then running the numbers through a computer's random selection program until 800 numbers are picked. This program will be based on a lottery principle so that each digit from 0–9 is found an equal number of times with no more repetitiveness than should properly occur by chance and with no tendency for the numbers to form repeating patterns. This process is usually carried out 'without replacement' so that once a unit has been selected it is not replaced into the population and therefore it cannot be chosen again.

'Simple' random sampling is particularly suitable to use where the population is relatively small and the sampling frame is complete. Managers can use it within their organisations and for some aspects of market research. It is 'simple' as compared with more complex methods and because it is based in a very straightforward way on the principle of random selection. Where there is a large population or an incomplete sampling frame, less simple methods may be used.

Systematic random sampling

This is a more complex form of random sampling because it involves a system: a system based on regularity. The sampling frame is taken and a name or unit is chosen at random, then from this chosen name or unit every nth item is selected throughout the frame. In other words there are regular intervals between the chosen names. For example, if the sampling frame consists of a company's wage list, there may be 10,000 names on it. If a 5 per cent sample is required, the 500 names can be selected at regular intervals. The first name is selected at random from the first 20 on the list (20 because $20 \times 500 = 10{,}000$). If the 12th name is picked at random, the names are then selected at regular intervals up to 500 in all (12th, 32nd, 52nd, 72nd, 92nd, 112th, etc).

This system facilitates the selection of sampling units where a well-organised list or sampling frame exists. At the same time, it is sufficiently random to obtain an estimate of the sampling error. On the other hand, it is not fully random, and is sometimes referred to as quasi-random sampling, because after the first point is selected randomly every remaining unit is selected by the fixed interval. Managers need to be aware that there can be problems if particular characteristics arise in the sampling frame at regular intervals. At one extreme, it would be possible for the whole sample to be based on these 'extreme' units. For example, if the sampling frame is based on the work position of employees it might be the case that because of the design of the factory or office, every 20th workplace is filled by a supervisor. The sample would be based on supervisors, which might not be the intention, and would cause bias in the results.

Providing that care is taken into investigation of this problem, this form of sampling can be useful in management where good sampling frames often do exist and also in marketing research based on the electoral register or a list of houses.

Stratified random sampling

This form of sampling takes the main problem of systematic

sampling and makes it into a virtue. All the people or items in the sampling frame are divided into groups or categories which are virtually exclusive. This means that a person or unit can be in one group or 'stratum' only. Within each of these strata a simple random or systematic sample is selected, the results of these samples are processed and if the same proportion of each stratum is taken, then they will be represented in the correct proportion in the overall result.

For example, in a survey on the labour force's views of changing working hours, employees may be divided into strata according to their roles in the company. A systematic sample of, say, 10 per cent can be taken within each stratum to arrive at the views of each group and the views of the whole labour force.

This method of sampling eliminates differences between strata from the sampling error which can be a problem in systematic sampling. It is useful where there are clearly defined groups within a population.

Random route sampling

This is a form of systematic sampling used particularly in marketing research surveys. The sampling frame is usually a list of properties (such as the addresses in the electoral register) and consists of houses, shops, garages, factories and offices. An address is selected at random as a starting point and the interviewer is then given instructions to identify further addresses by taking alternate left- and right-hand turns at road junctions and calling at every nth address 'en route'. The people at these addresses are then interviewed or given questionnaires to complete.

This is a quasi-random method of sampling because the element of selection can be strong although the interviewer has to call at clearly defined addresses and is not able to choose. Geographical areas have to be chosen carefully because they have particular characteristics. In carrying out marketing research on consumer goods, a company may decide to use random route sampling just because it is possible to choose a particular type of area, a suburb for example, which has just the correct characteristics for the consumers being targeted.

It is difficult to check that interviewers have carried out the sample correctly without a costly repetition of the route. Addresses may have been missed, non-response not recorded or certain premises avoided because they did not look very promising from the interviewer's point of view. Managers can make spot checks to reduce this problem, and can provide training for the interviewers, and above all they should be aware of the problems that can arise with research based on this commonly used method.

Quota sampling

This is another commonly used method of carrying out marketing research and opinion polls. In quota sampling the interviewer is instructed to interview a certain number of people with specific characteristics. The quotas are chosen so that the overall sample will reflect accurately the known population characteristics in a number of respects. It is not a random sampling method because every individual does not have a known probability of being included in the sample. Quota sampling can be described as non-random but representative stratified sampling. It can be used when there is not a suitable sampling frame to use a random sampling method.

A marketing research company may be asked to discover the effects of a promotional campaign on the targeted consumer group. These may be families in the high income bracket. The interviewers can be sent to town centre shopping areas and told to interview over a period of a few days twenty people between 40 and 50 years of age and twenty between 50 and 60, all of whom are in families and earn an above average income. The interviewers have to try to pick people falling into these characteristics from the shoppers and then ask them the questions that have been provided. These will usually start by checking the characteristics: 'What is your date of birth?/Do you have a family?/Is your family income above the average national wage?'

The more characteristics that are introduced, the more difficult it is for the interviewers to select the respondents and interviewer bias may become an important factor. Some people may be over-represented, for example people who happen to be out shopping on particular days or/and people who 'look' as if they are articulate and answer questions easily.

Non-response may not be recorded because people who refuse to be interviewed are not necessarily recorded in any way.

On the other hand this is a method which can be introduced and carried out quickly and relatively cheaply.

Cluster sampling

This is a geographical or 'area' method of sampling. Clusters are framed by breaking down the area to be surveyed into smaller areas. A few of these areas are then selected by random methods and individuals or whole households are interviewed in these selected areas. The people within the areas may then be selected by random methods.

The sampling frame will usually be a map and an area of this (such as a town) is divided into sections. A random selection is made of these sections and every household can be interviewed, or systematic or random route sampling used. A team of interviewers is then sent from one section to the next so that the survey can be carried out quickly.

Where there is an inadequate sampling frame in the form of a list of names or addresses, this geographical method may be the only possible method. The clusters of people surveyed may have similar characteristics and therefore the result may be biased. All the sections chosen, for example, may be relatively poor; the chances of this can be reduced by taking a large number of small samples. This 'cluster' sampling is often used to survey the distribution and possible markets for consumer durables such as refrigerators and video-players. Also it is used for quality control in manufacturing companies. Random batches of items are removed from the production line and all units within the batch are tested.

Multi-stage sampling

This describes the way a sample survey is organised, rather than any fundamental sampling method. Multi-phase sampling, replicated and interpenetrating sampling and master samples are similarly forms of sample design rather than methods.

Multi-stage sampling is a system of taking a series of samples at successive stages, while in multi-phase sampling some information is collected from the whole sample and additional information is collected from sub-samples of this. Replicated or interpenetrating sampling is a design which involves a number of sub-samples, rather than one full sample, selected from a population. A master sample covers the whole country to form the basis or provide a sampling frame for smaller, local samples.

These methods are used to sample very large populations such as the whole population of a country. In multi-stage sampling the country may be divided into geographical regions, a limited number of towns and rural areas are then selected in each region and samples are taken of people or households in these towns and areas. The aim is to select towns and rural areas in such a way that the probability of selection is proportional to the size of the population. A town with a population of 100,000 would stand ten times the chance of being selected as a town with a population of 10,000 and an individual in the large town would stand one-tenth the chance of being selected as compared with an individual in the smaller town, so that individuals in both towns stand an equal chance of selection at the beginning of the sampling operations.

The UK Family Expenditure Survey, for example, uses multi-stage sampling methods and forms the basis of the Index of Retail Prices and the measurement of levels of inflation. The general approach is:

- i the country is divided into 1,800 areas, from which a stratified (by region, rural and urban, and level of community charge) sample of 168 areas is chosen
- ii the selected areas are divided into districts and four districts from each are selected by systematic sampling
- iii 16 households are selected from each district by taking a systematic sample from the electoral register.

This produces a sample of approximately 10,000 households (16 households × 4 districts × 168 areas = 10,752 households in total). Some households do not co-operate, but most do and every member of these selected households over 16 years of age is surveyed.

Managers may use replicated or interpenetrating samples within their organisations, particularly where there are large numbers of employees. A 10 per cent sample of 20,000 employees may be divided into two sub-samples of 1,000 employees each, or four sub-samples of 500 employees. The smaller samples can be processed very quickly to provide early results, while the many sub-samples can show up non-sampling errors because their design is exactly the same and each is a self-contained sample of the population. It is also possible to obtain an estimate of variations between interviewers if each sample is carried out by a different interviewer.

Panels

These are samples in the sense that a group of people is selected from a survey population by a random process. This group forms a panel of people who are questioned for information at various intervals over a period of time. This means that the same information is asked for from the same sample at different times.

A marketing manager may use this to test the success of a promotion campaign. The panel will be asked about the product before the campaign and then will be asked similar questions after the campaign, and the results are then compared.

Panels are often used to measure trends in behaviour and changes in opinions. The main problems arise out of people leaving the panel or the panel becoming untypical because of 'panel conditioning'. This develops because the members may, over a period of time, become untypical of the population they represent. A panel of members in a company may be asked questions about the company organisation over several months. They may, however, become more interested in the organisation and structure than most members of the labour force and therefore no longer be typical.

Marketing research

The most obvious application of sampling methods in management is in marketing research. If there is a difference between *market* research and *marketing* research, it is that market research is concerned with the measurement and analysis of markets, whereas marketing research is concerned with all those factors which impinge upon the marketing of goods and services. Most so-called market research is in fact concerned with marketing rather than markets.

Marketing research is the planned, systematic collection, collation and analysis of data designed to help the management of an organisation to reach decisions and to monitor the results of these decisions. Essentially, it seeks to provide answers to five basic questions: Who? What? When? Where? and How? and also perhaps even Why? These are summarised in Table 4.2. All

Table 4.2 *Marketing research questions*

A manager wants to know:
Who buys our products/services?
What products/services do people want?
What benefits are they looking for?
When will people buy our products/services?
Where will they buy them?
How will they pay?
Why will they buy our products/services and not our competitors'?
What can we do to make our products/services attractive?

organisations need to know the effectiveness of their advertising and promotion campaigns, they want fully to understand the customer and to answer the five or six basic questions about them. When new products or services are planned, the organisation needs to research the marketing opportunities. Marketing research can help to reduce uncertainty concerning the outcome of future events or a given course of action. The process is usually as shown in Figure 4.1.

Marketing research techniques are those of survey design and sampling methods. There are also other methods used by management in practice. These may be a preliminary to a probability sample, 'quick and dirty' methods used in order to clarify basic issues and to provide generalised information in a hurry. These techniques include:

Convenience sampling: this consists of soliciting information from any convenient group whose views may be relevant to the subject of the inquiry. For example, shoppers may be asked their views on a new service to obtain a feel for the subject and as a basis for formulating more precise questions to be asked of a representative sample. Companies may ask their staff to sample a new product. They may be biased because their jobs depend on the product but on the other hand they will also want to be realistic about the possible success of the product.

Snowball sampling: participants in convenience sampling could be asked to suggest the names of others 'like them' who could be contacted. This would add a group that does not have the same biases as the first group (it is possible to insist that they are not employed by the organisation).

Piggybacking: is adding questions to an existing survey or using

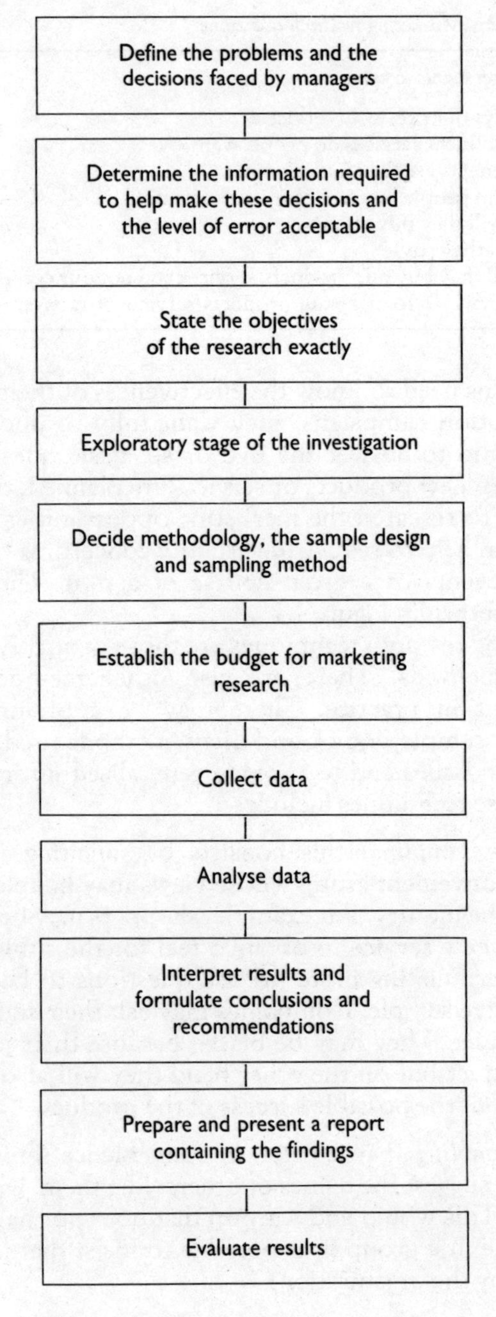

Figure 4.1 *The marketing research process*

an established form of distribution in order to conduct a survey. For example, a questionnaire can be added to a mailing to magazine subscribers, or interviews may be carried out at a meeting arranged for a different purpose.

Focus group interviewing: consumers are interviewed in groups of say ten targeted consumers, usually a relatively homogeneous group brought together to discuss specific issues under the guidance of a leader trained to stimulate and focus the discussion.

Judgement sampling: respondents are selected on the basis of the interviewer's subjective opinion so that they constitute a representative cross-section of the population to be investigated.

These are not random sampling methods but for managers they can provide an important element in the exploratory stage of an investigation and can raise pointers towards the type of sample method and survey method to be used.

Experiments

Managers may collect information through experiments. In the 'ideal' situation of the laboratory experiment, the environment is controlled and the effect of one variable on another can be studied in isolation from extraneous influences. This is more difficult in human and social sciences than in physical science, but statistical randomisation can be used in order to obtain the safeguards it provides in reducing chance factors.

An experiment involves the manipulation of one variable to see what effect this may have on another variable. Phenomena are carefully measured and the data used to test hypotheses which attempt to explain what is happening. The results of experiments can be used to make statements about general tendencies or 'laws' which can be used for 'prediction' and forecasting.

In statistical experiments randomisation can be used in order to provide the safeguards it produces in reducing chance factors. Two groups of people can be 'matched' for factors such as age and sex which might affect their attitudes and be chosen randomly so that differences between them can be measured. In the 'before–after' experimental design, for example, a matched 'control' group can be used so that when the 'experimental' group is exposed to a pilot promotional campaign the control group is not and the resulting changes in attitude can be compared between the two groups. The results might show, for example, that there has been a change in attitude or at least awareness in the experimental group which is greater than any change in the control group. This might lead to a manager reaching the conclusion that the promotional campaign should go ahead. This is, however, a matter of judgement and the experiment will be simply one factor in reaching a decision.

Experiments will tend to be used where the results need to be very accurate, the area of study is narrow and the various factors involved can be controlled. Test marketing of a new product might be carried out on the basis of an experimental design. Surveys and samples will be used where the margin of error that can be tolerated can be wide and where the results can be generalised to other people and they tend to be used for general marketing research and for opinion polls.

- A *sample* is anything less than a full survey of a population.
- A *population* is the group of people or items about which information is collected.
- The *law of statistical regularity* states that a reasonably large sample selected at random from a large population will be, on average, representative of the characteristics of the population.
- The *law of the inertia of large numbers* states that large groups of data show a higher degree of stability than small ones.
- *Sample size* does not depend on the size of the population; it depends on the resources available and the accuracy required.
- *Sampling errors* arise because a sample cannot exactly represent the population from which it is chosen.
- *Sample design* depends on the objectives to be achieved and the resources available.
- The *sample frame* is a list of people, items or units from which the sample is taken.
- *Randomness* means that every item in a population has an equal chance of being included in the sample.
- *Sampling methods* include:

  Random methods: simple random sampling
systematic random sampling
stratified random sampling
random route sampling

Non-random methods: quota sampling
cluster sampling
multi-stage sampling

- *Market research* is concerned with the measurement and analysis of markets.
  Marketing research is the planned, systematic collection, collation and analysis of data designed to help the management of an organisation to reach decisions with regard to the marketing of goods and services and to monitor the results of these decisions.

Part 2
Statistical Analysis for Managers

5
The Analysis of Management Information

The need to analyse data in management

Managers receive large quantities of statistical data through internal reports, monitoring procedures, quality control, production figures, sales forecasts, marketing research findings, wage returns and so on. They may be concerned also with regional, national or international comparisons, import/export figures, balance of payments, and financial returns. In making use of these data, understanding them and communicating the salient facts to other people, managers need to analyse them and summarise them.

Summarising data is an important part of analysis because the summary can provide an understanding of groups of figures and their possible relationship to others. If pay returns in a company show that the average salary paid to secretaries is 15 per cent lower than in other organisations in the area this may be a good indication of why the company has difficulty in recruiting secretaries. In applying quality control to a production line, the average size of a component may be agreed and deviation from this average monitored, so that those units which deviate from the average beyond a level that can be tolerated can be rejected.

Both averages and measures of deviation from them are methods of summarising and analysing data. These measures can be arrived at very easily by instructing a computer to produce them at the same time as it produces the base data. It still remains, however, very important for managers to understand the measures with which they are presented even if they do not have to calculate them. This is the only way that a manager will know when to use one average rather than another, or will appreciate what is meant when a particular measure is included in a report.

There are complex inductive statistical methods of analysis which a manager may come across. These include techniques such as the analysis of variance, regression, multiple and partial correlation, and factor analysis. These methods require a mathematical background beyond the scope of this book and it is unlikely that many managers will need to work with them. If any of these measures do become important, then the background work will need to be carried out just as it is with the more common descriptive methods described here.

The role of the average

Averages are measures of central tendency and measures of location. As a measure of central tendency an average provides a value around which a set of data is located. As a measure of location an average provides an indication of whereabouts the data are situated. A manager who is told the average pay of a group of employees will have some idea whether an employee within the group will be paid £500, £1,000 or £3,000 a month.

An average summarises a group of data and represents it in the sense that the average provides an immediate idea about the group; it can provide a description of a group of items so as to distinguish it from another group with similar characteristics and it can do this concisely. In comparing two groups of employees the fact that one group receives average pay twice as high as the other group provides an immediate and concise distinguishing feature. The average used depends on the type of description required. The three most commonly used are the arithmetic mean, the median and the mode.

These averages are used all the time in everyday life and work although they are not necessarily used accurately. Managers refer to the 'average cost, average price, the average wage' and also to the 'usual size' and the 'normal price'. These may be arithmetic means (average cost, average price), or medians (average wage), or modes (usual size, normal price). A little more information would make this clear.

Averages can:

i Summarise a group of figures, smoothing out extremes in a way that is useful for comparison. For example: a salesman may have sold 5,200 units over a year although in some weeks he sold two units and other weeks, two hundred. His average (arithmetic mean) sales of 100 units per week is an arithmetic mean and this figure can be compared with the performance of other salesmen.

ii Provide a mental picture of the distribution it represents. For example: a holiday company may have an average monthly income of £20,000 and this provides an immediate idea of the size of the business, although in the good months it is £80,000 and in the poor months it is £4,000.

iii Provide valuable knowledge about the whole distribution. A manager who knows that there are 100 employees in a company with average monthly pay of £750 will be able to calculate that the total monthly wage and salary bill will be £75,000 and the annual bill £900,000.

iv Conceal important facts if they are the only piece of information available. Although the average monthly pay may be calculated to be £750, the actual pay may be distributed very unevenly, so that most employees are paid between £400 and £500 a month with a relatively few paid between £3,000 and £5,000 a month.

v Provide the first stage of an investigation and are often used as an estimate. A saleman may say 'on average I drive about 600 miles a week' as an estimate. Further investigation of the mileage claimed over a year may confirm this figure or show that it was a very approximate guess.

General characteristics of the arithmetic mean, the median and the mode

The arithmetic mean is the average to which most managers refer or think they refer when they use the word 'average'. It can be defined as the sum of the items divided by the number of them (£100 divided equally between four people gives an arithmetic mean of £25). This is widely understood by managers and the basic calculation is straightforward.

There are times when people may think they are referring to the arithmetic mean when they are not. Statements such as: 'The average pair of shoes costs £20', are more likely to be a description of the mode than of the arithmetic mean. If the statement was that: '*Most* pairs of shoes cost £20', this would be an accurate definition of the mode. The mode is the most frequently occurring value in a distribution and it may be that the most frequent price for a pair of shoes is £20. The mode is often used in general conversation and at work because it does often represent a typical item and may appear more realistic and sensible than the arithmetic mean. It is not, however, necessarily well defined and may ignore important aspects of distributions which are better represented by the arithmetic mean.

The median may be useful at times, particularly where extreme values can provide a distorted summary of a distribution. The median is the value of the middle item of a distribution which is set out in order. If five people earn £300, £350, £380, £420, £1,550 a month respectively, the median is the value of the middle item, £380. This could be considered to be a fair representation of most of the items, and it is closer to most of the items than the arithmetic mean of £600 (£3,000 divided by 5). It is useful if managers know which averages they are using, particularly if statements about averages are likely to be followed up in more detail.

The arithmetic mean

This can be defined as the sum of the items divided by the number of them:

$$\text{The arithmetic mean} = \frac{\text{total value of items}}{\text{total number of items}}$$

If ten people have bonus payments of £60, £70, £90, £100, £102, £107, £110, £115, £116 and £130 respectively, the arithmetic mean will be the total value of these payments (£1,000) divided among ten people: £100 each. This means that if all bonus payments had been shared equally among the ten people, they would have received £100 each.

$$\text{The arithmetic mean} = \frac{£1{,}000}{10} = £100.$$

This can be written as:

$$\text{The arithmetic mean} = \frac{\Sigma x}{n}$$

Where (sigma) Σ = the sum of (or the total of)
x = the value of the items
n = the number of items.

The usual symbol for the arithmetic mean is $\bar{x}$ (x-bar or bar-x), although strictly speaking this is the symbol for the arithmetic mean of a sample and μ ('mu' or 'mew') is the symbol for the arithmetic mean of the population from which samples are selected. The alternative $\bar{x}$ is very widely used for the arithmetic mean and will be used here.

Another method of calculating the arithmetic mean, by using

Table 5.1 *Arithmetic mean*

Bonus (£)	*Difference from assumed mean (£105)*
60	−45
70	−35
90	−15
100	−5
102	−3
107	+2
110	+5
115	+10
116	+11
130	+25
	−103 + 53 = −50

$$\bar{x} = £105 - \frac{50}{10}$$
$$= £105 - 5$$
$$= £100.$$

deviations from an assumed mean, is useful to understand because it is the basis for calculations with grouped data. In this method an average is guessed at (or assumed) by inspection, the differences (or deviations) of the items from this guess are found and the sum of these differences is divided by the number of items and added or subtracted from the guess.

Using the same figures as before, Table 5.1 shows the result if an arithmetic mean of £105 is guessed. There are two points to note here. One is that any figure chosen as the assumed mean would give the correct answer (try it!). The other is that the formula used here is:

$$\bar{x} = x \pm \frac{\Sigma d_x}{n}$$

where x represents the assumed mean
d_x represents the differences
(or deviations from the assumed mean)
Σd_x is the sum of these differences
n is the number of items.

In this case £5 is subtracted from the assumed mean to give the same answer as before.

The arithmetic mean of a grouped frequency distribution

Managers will be presented from time to time with tables of grouped frequency distributions, in which the items are not discrete but are classified in groups and within each group there are varying frequencies. The calculation of the arithmetic mean is based on the assumption that the frequency equals the mid-point of the class interval because there is no way of knowing the actual distribution of frequencies within a class.

There are two methods of calculation, one concentrating on the mid-point of the class and the other on the class interval.

The mid-point method

The problems of classification have been covered in Chapter 2.

Table 5.2 *The mid-point method*

(1)	(2)	(3)	(4)	(5)
Weekly Wages (£)	*Number of employees (f)*	*Mid-point of class interval (m.p.)*	*Deviation of mid-point from assumed mean (£145) (d_x)*	*Frequency $\times d_x$ (2) × (4) (Σfd_x)*
120 but less than 130	5	125	−20	−100
130 but less than 140	12	135	−10	−120
140 but less than 150	18	145	0	0
150 but less than 160	45	155	+10	+450
160 but less than 170	20	165	+20	+400
	100			−220 +850
				$\Sigma fd_x = +£630$

$$\bar{x} = x \pm \frac{\Sigma fd_x}{\Sigma f}$$

$$= £145 + \frac{630}{100}$$

$$= £145 + £6.30$$

$$= £151.30.$$

Where x is the assumed mean
f is the frequency
d_x is the deviation of the mid-point from the assumed mean

The mid-point of a class can be found by commonsense observation or by adding the lower limit of one class to the lower limit of the next class interval and dividing by two:

$$\frac{140+150}{2}=\frac{290}{2}=145$$

The class interval method

Instead of using the mid-point of the class intervals this method is carried out 'in units of the class interval'. The deviation of classes from the assumed mean is in units of 10 in this case.

The methods produce the same result (!) and it does not matter which is used. The class interval method may be quicker in some circumstances while on the other hand it may be easier to use the mid-point method when there are uneven class intervals.

Table 5.3 *The class interval method*

Weekly wages (£)	*Number of employees (f)*	*Deviation of classes from class of the assumed mean* (d_x)	*Frequency* (fd_x)
120 but less than 130	5	−2	−10
130 but less than 140	12	−1	−12
140 but less than 150	18	0	0
150 but less than 160	45	+1	+45
160 but less than 170	20	+2	+40
	100		−22 +85
			$\Sigma fd_x = +£63$

$$\bar{x}=x\pm\frac{\Sigma fd_x}{\Sigma f}\times\text{class interval}$$

$$=£145+\frac{63}{100}\times 10$$

$$=£145+0.63\times 10$$

$$=£145+6.30$$

$$=£151.30$$

The median

The median can be defined as the middle item of a distribution which is set out in order. If five employees have monthly earnings of £600, £750, £900, £1,000 £2,500, the median is the value of the earnings of the employee in the middle of the five: £900.

i *A discrete distribution*: in a discrete distribution such as the one above, the median can be ascertained by inspection. The formula for finding the median position for a discrete series is:

$$\text{The median} = \frac{n+1}{2}$$

where n is the number of items.

In the example, this would be:

$$\frac{5+1}{2} = 3.$$

The median is the value of the third item (£900). If there is an even number of items, the values of the middle items are added together and divided by 2: if six employees have monthly earnings of £600, £750, £900, £1,000, £1,200, £2,500,

$$\text{then the median position} = \frac{n+1}{2}$$

$$= \frac{6+1}{2}$$

$$= 3.5.$$

$$\text{The median would be:} \quad \frac{£900 + £1{,}000}{2}$$

$$= \frac{£1{,}900}{2}$$

$$= £950.$$

The usual symbol for the median is M. These examples show that the median is unaffected by extreme values (such as £2,500), and this is one reason for using the median as an average in particular circumstances. Personnel managers often use it when analysing wage and salary distributions because many employees may have pay which is close to each other's while a few may be at an extreme with relatively low or high pay. At the same time the median divides a distribution in half by the number of items, not their values, and this may be useful with such lists as those for wages. Used in conjunction with the upper and lower quantities, the median can divide a pay list into four equal parts, which is helpful in comparing trends or comparisons with other groups of employees.

ii *Grouped frequency distribution*: in this type of distribution, the median can be found by two methods: (a) by calculation and (b) graphically.

Calculation of the median

Table 5.4 *The median*

Monthly pay (£)	*Number of employees*	*Cumulative frequency*
500 but less than 600	5	5 employees receive less than £600
600 but less than 700	15	20 employees receive less than £700
700 but less than 800	20	40 employees receive less than £800
800 but less than 900	50	90 employees receive less than £900
900 but less than 1,000	10	100 employees receive less than £1,000
	100	

The position of the median for a continuous series is:

$$\frac{f}{2}$$

where f is the total frequency. In Table 5.4 the position of the median is:

$$\frac{100}{2} = 50.$$

The 50th employee's salary is the median pay. This falls within the class interval £800 but less than £900. The first 40 employees in order earn up to £800 a month; the 50th employee earns at least £800, but less than £900.

The only way to arrive at a closer calculation of the salary of this employee is to assume that the earnings are evenly distributed within the class interval. The 50th employee is the tenth person in the group earning between £800 and £900 a month because the previous group includes the 40th employee and $50-40=10$.

If it is assumed that the £100 in the class interval of this grouped data is divided evenly between the 50 employees in the group, then the 50th employee's salary will be 10/50ths of £100 plus the starting point of £800:

$$\begin{aligned} M &= £800 + \frac{10}{50} \times 100 \\ &= £820. \end{aligned}$$

The median monthly salary is that of the 50th employee which is £820. This divides the distribution in half, so that of the 100 employees half will earn less than £820 and half will earn more. A manager may be able to compare this situation with a similar group of employees. If another company, for example, were shown to have a median monthly salary of £1,200 for a very comparable group of employees, this might explain difficulties in recruitment in the manager's company.

It should be noted that it is only the class interval of the class in which the median falls that is important; irregular class intervals in a grouped distribution are not a problem because they can be ignored.

The median found graphically

The median is found graphically by drawing the cumulative frequency curve or ogive. 'Ogive' is an architectural term which is used to describe an S (or flattened S) shape, similar to the usual shape of the cumulative frequency.

Using the data in Table 5.4, Figure 5.1 shows the cumulative frequency plotted against monthly salaries. It is important to notice that in Figure 5.1, when plotting the cumulative frequencies, the point must be placed at the end of the group intervals.

In Table 5.4:

$$\text{The position of } M = \frac{100}{2} = 50.$$

A horizontal line is drawn from this point (50) on the vertical axis to the ogive, and at the point of intersection a vertical line is

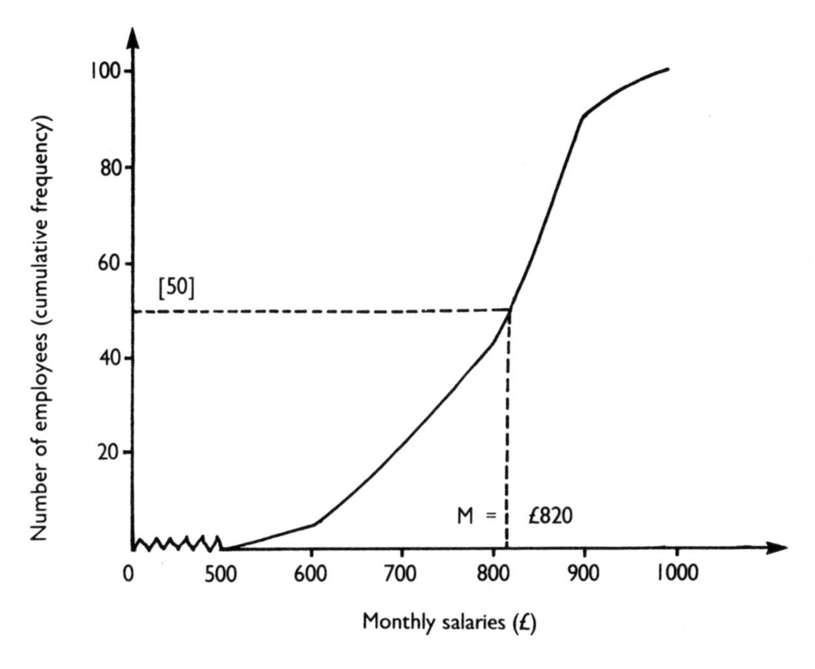

Figure 5.1 *Cumulative frequency curve*

dropped to the horizontal axis. This shows that the value of the median is approximately £820. The more accurately the graph is drawn, the more accurately it is possible to interpolate the median.

The median divides an ordered distribution into half, and in a similar way it is possible to divide distributions in quarters (quartiles), tenths (deciles) and so on. The method of calculation or interpolation is similar to that of the median. This will be considered, with the question of dispersion, in the next chapter.

The mode

The mode can be defined as *the most frequently occurring value in a distribution*. It is often used in the sense of 'most' or 'usual' or 'normal': 'most' families, the 'usual' income, the 'normal' size. It may be stated that the average employee has an income of £8,000; when the arithmetic mean is calculated it may be discovered that it is £12,000 because of a few particularly high salaries, although 'most' people earn £8,000. This is the most frequently occurring value, or mode. It is the 'normal' or 'usual' income earned by this group of employees.

In a discrete list of figures the 'mode' is the most frequently occurring number: £7,000, £8,000, £8,000, £9,000, £12,000, £28,000 (the mode is £8,000).

Table 5.5 *Frequency distribution*

Salaries (£)	*Number of employees earning these salaries*
5,000	10
6,000	15
7,000	20
8,000	40
9,000	15

In a frequency distribution such as Table 5.5, the mode is the item with the highest frequency. The mode is £8,000, because 40 is the highest frequency (£8,000 is the most frequently earned salary).

Calculating the mode for a grouped frequency distribution is not easy or particularly useful, because since a grouped frequency distribution does not have individual values it is impossible to determine which value occurs most frequently. At the same time, if different sets of class intervals are drawn when the classifications are being decided, the modal class might be different or might include different values. The modal class is the one with the highest value and this can be useful as a form of description.

The modal class in Table 5.6 is '4 cm but less than 6 cm' because this class has the highest frequency (24). If the classification had been different, say:

Up to 4	13
4 but less than 8	36
8 but less than 12	11

the modal class would have been '4 but less than 8'. Although this would have produced a different result from the same distribution, the modal class would still have been around the centre because this is a unimodal distribution which is not

Table 5.6 *Modal class*

In an assembly plant components are used in the following quantities:

Components (cm)	*Number used*
Up to 2	3
2 but less than 4	10
4 but less than 6	24
6 but less than 8	12
8 but less than 10	9
10 but less than 12	2
	60

Figure 5.2 *The mode*

heavily skewed (see Chapter 6). Where the distribution is heavily skewed or 'multi-modal' the use of the modal class as a description becomes less useful.

The modal class can be illustrated by drawing a histogram

(Figure 5.2) and from this it is possible to estimate the mode. This estimation is carried out by drawing a line from the top right-hand corner of the modal class rectangle or block to the point where the top of the next adjacent rectange to the left meets it (line *x* to *y* in Figure 5.2); and a corresponding line from the left-hand top corner of the modal class rectangle to the top of the class on the right (line *a* to *b*). Where the two lines cross a vertical line can be dropped to the horizontal axis and this will show the value of the mode. Using the data in Table 5.6 to draw the histogram in Figure 5.2, the mode can be seen to be approximately 5 cm. This can only be an estimate, because the individual values are not known, but it may provide a useful general summary of a distribution.

- *Averages* are measures of central tendency and measures of location.
- *The arithmetic mean* is the sum of the items divided by the number of them.

$$\text{AM or } \bar{x} = \frac{\text{total value of items}}{\text{total number of items}}$$

$$= \frac{\Sigma x}{n}.$$

With a grouped frequency distribution:

i mid-point method

$$\bar{x} = x \pm \frac{\Sigma f d_x}{\Sigma f}$$

ii class interval method

$$\bar{x} = x \pm \frac{\Sigma f d_x}{\Sigma f} \times \text{class interval}$$

- The *median* is the middle item of a distribution which is set out in order.

The median (M) position $= \frac{n+1}{2}$ or $\frac{f}{2}$ (for a continuous series).

- The *mode* is the most frequently occurring value in a distribution.

6
Further Analysis of Management Information

The role of dispersion

Measures of dispersion are methods of summarising and comparing data; they can be described also as measures of deviation or spread. Distributions are not only clustered around a central point or average, they are also spread around it. Averages do not provide an idea of the form or shape of a distribution while measures of dispersion can do this.

Distributions can be a great variety of shapes when they are plotted on a graph. Some of these shapes are common to many distributions, at least to an approximate extent.

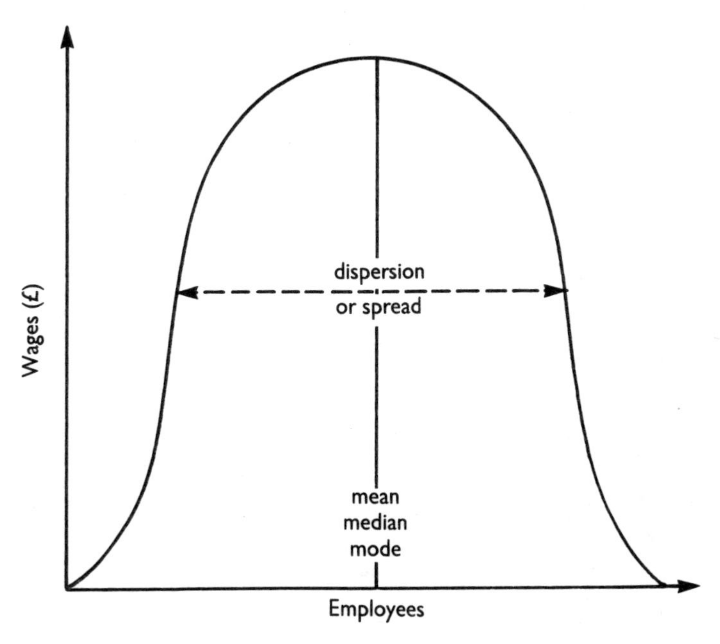

Figure 6.1 *Bell-shaped distribution*

Distributions

i A bell-shaped or 'normal' curve (Figure 6.1) is symmetrical in that the distribution is spread equally on either side of the arithmetic mean, median and mode. Biological data such as height and weight are often distributed in this way because most people are of 'average' height or weight with some at various points below the average and some above. Wage distribution will approximate to this shape if most people in a company are paid close to a median wage with those paid less balanced by those paid more.

ii Many distributions are 'skewed' (Figure 6.2), so that the peak of the curve on the graph is displaced to the left or the right. When the peak is displaced to the left of centre, the distribution is described as positively skewed; when the peak is displaced to the right, the distribution is negatively skewed. Prices are often distributed in this way, with most brand prices of a commodity clustering around, say, a relatively low bracket while a few examples of the commodity are more highly priced or vice versa.

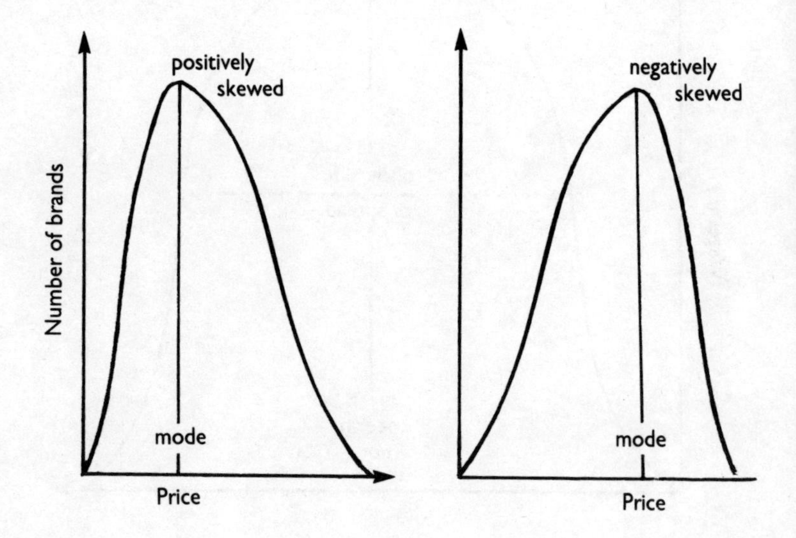

Figure 6.2 *Skewed distributions*

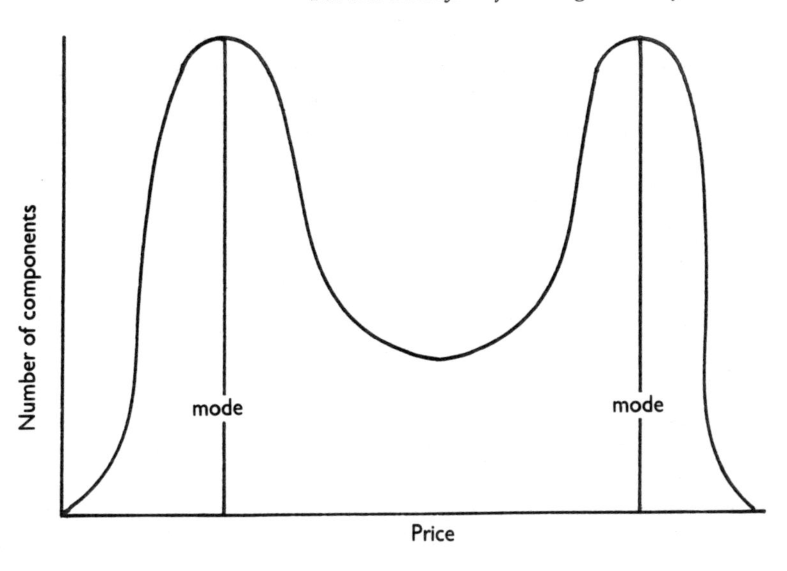

Figure 6.3 *Bi-modal distributions*

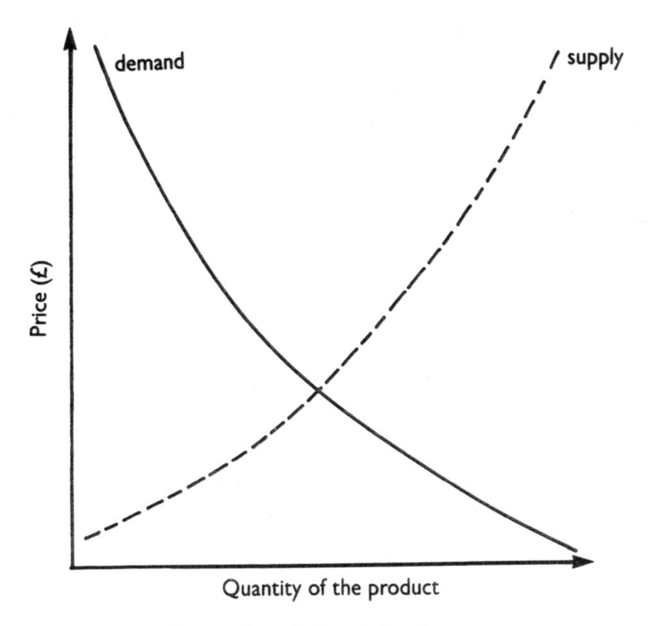

Figure 6.4 *J-shaped distribution*

iii Some distributions are bi-modal (Figure 6.3), in that there are two peaks rather than one. This may arise with spare parts, for example, where there are highly priced parts and low-priced, budget equivalents, without much in between.

iv The J-shaped distribution or reverse J-shaped (Figure 6.4) are typical of supply and demand graphs. The demand for a product or service is likely to be greater at a lower price than at a higher price, so that the curve will slope down from left to right. The supply curve is the opposite, with the willingness to supply greater when the price is high than when it is low.

There are, of course, many other shapes that distributions outline once they are graphed and they can, usually with the help of an average, provide a short description of the data. As well as showing the spread of a distribution diagrammatically, it is also possible to calculate a measure of dispersion or deviation, the simplest of which is the range (see below).

The Lorenz curve

This is a graphical method of showing the deviation from the average of a group of data. It can be described as a 'cumulative percentage curve' and it is used to show the inequalities between two variables. Cumulative frequencies are calculated from the basic data and are then compared on a graph. The less equal the distributions, the wider the curve, while equality between the two variables would produce a straight line: the line of equal distribution.

Table 6.1 and Figure 6.5 illustrate this method.

Table 6.1 *Company size and capital investment*

(1) *Revenue (£m)*	(2) *No. of companies (f)*	(3) *Capital investment (£m)*	(4) *Cumulative capital investment (£m)*	(5) *(4) as %*	(6) *Cumulative frequency of (2)*	(7) *(6) as %*
Less than 2	200	36	36	18	200	50
2 but less than 4	100	24	60	30	300	75
4 but less than 6	65	40	100	50	365	91.25
6 but less than 8	20	44	144	72	385	96.25
8 but less than 10	15	56	200	100	400	100

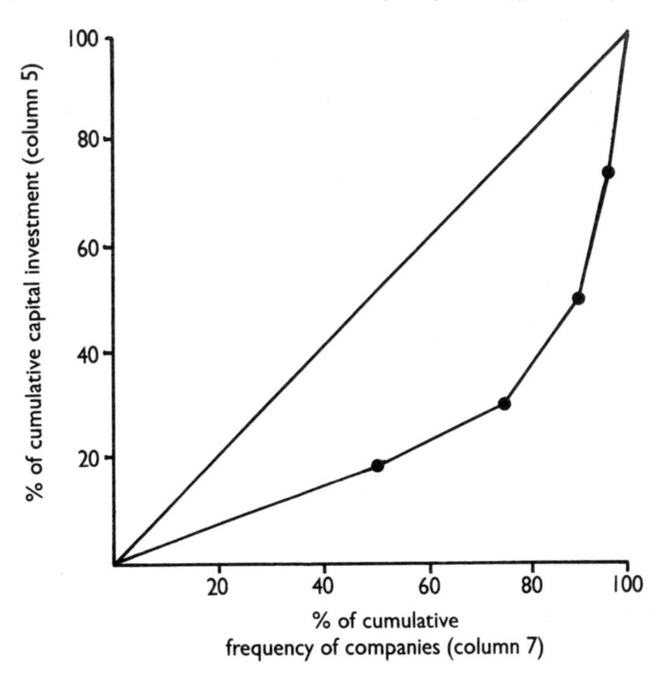

Figure 6.5 *Lorenz curve*

The range

This is a commonly used method of describing the dispersion of data. It is defined as 'the highest value in a distribution minus the lowest'. Managers make frequent use of this measure when discussing prices, wages, sizes, and so on. 'Our car prices start at £7,000, with the top of the range at £15,500', for example. The range in fact is £8,500: £15,500 – £7,000. Another car dealer may have a narrower range, say £5,000: £6,000 to £11,000.

The range is often used with the mode in order to provide a clearer picture of a distribution. Although, for example, car prices may range from £7,000 to £15,500 there needs to be some knowledge of an average price to give an indication of whether most of them are priced close to £7,000 or £15,500 or are spread evenly through the range. If it is known that the 'usual' or 'most popular' cars sold are priced at £9,000, this indicates that the £15,500 cars are expensive and rarely bought and that the £7,000 cars are relatively cheap. In this case the average used is the mode ('the most frequently occurring value').

The interquartile range

This is a very similar measure of dispersion to the range but it measures a different part of a distribution. The interquartile range can be defined as the difference between the upper quartile and the lower quartile ($Q_3 - Q_1$).

The median divides an ordered distribution into half, and in a similar way it is possible to divide distributions into quarters. This divides an ordered distribution into four equal parts by calculating 'three' quartiles: the lower quartile (Q_1), the middle quartile (Q_2 or M), which is the same as the median, and the upper quartile (Q_3).

In a distribution of 100 items the quartiles will be the values of the 25th (Q_1) and the 75th (Q_3) items, with the median the value of the 50th item. The method of calculating the lower and upper quartiles is very similar to the method for calculating the median.

Table 6.2 *The quartiles*

Monthly pay (£)	*Number of employees*	*Cumulative frequency*
500 but less than 600	5	5 employees receive less than £600
600 but less than 700	15	20 employees receive less than £700
700 but less than 800	20	40 employees receive less than £800
800 but less than 900	50	90 employees receive less than £900
900 but less than 1,000	10	100 employees receive less than £1,000
	100	

The position of the lower quartile $(Q_1) = \frac{n}{4}$, in Table 6.2.

This will be:

$$\frac{100}{4} = 25.$$

The pay received by the 25th employee will be the lower quartile pay. This lies in the class '£700 but less than £800', which is shared by 20 employees.

$$Q_1 = £700 + \frac{5}{20} \times 100$$
$$= £725.$$

The position of the upper quartile (Q_3) $= \frac{3n}{4}$, in Table 6.2.

This will be:

$$\frac{300}{4} = 75.$$

The pay received by the 75th employee will be the upper quartile pay. This lies in the class '£800 but less than £900', which is shared by 50 employees.

$$Q_3 = £800 + \frac{35}{50} \times 100$$

$$= £870.$$

The distribution can be divided in the following way:

$$\begin{aligned} Q_1 &= \text{25th employee earning £725} \\ M &= \text{50th employee earning £820} \\ Q_3 &= \text{75th employee earning £870.} \end{aligned}$$

The interquartile range will be:

$$Q_3 - Q_1 = £870 - £725 = £145.$$

It is now possible to describe this distribution by saying that the average (median) monthly pay is £820 with a range of £500 (£1,000–£500) and an interquartile range of £145. The interquartile range describes the pay of the middle half of the employees where there is a relatively small range compared with the full distribution. At the same time it can be seen that a quarter of the employees are earning less than £725 and a quarter more than £870.

The median and the quartiles are often used for summarising and analysing wage distributions and they can be compared easily with national statistics or with those for other companies. The interquartile range is not influenced by extreme items. For example, the top ten employees in Table 6.2 earn above £900. If they earned over £2,000 a month, the interquartile range would remain the same, between £725 and £870 (£145). Very high or very low earnings can 'distort' the description of a wage and salary list in giving a picture of the general level of pay in an organisation if only the range and the arithmetic mean are used and in these circumstances the interquartile range may provide a useful summary.

The standard deviation

Interpretation of the standard deviation

There is a marked difference between the standard deviation as a measure of dispersion and the range and the interquartile range. The last two both rely on the particular values within a distribution, the highest and the lowest values in the case of the range and the values of the upper and lower quartiles in the case of the interquartile range. This can lead to 'distortion' in the sense that it is possible that the two values are not very representative of all the values in a distribution. On the other hand they are useful 'everyday' measures for managers to use in analysing and summarising data.

The standard deviation is less useful in making a distribution more easily understood and its value in terms of communicating the properties of a distribution is likely to be greatest to a fairly specialised audience, that is, to those who already understand it. It is particularly important to some managers, however, because of its use in sampling theory and its applications in such areas as marketing research.

The standard deviation uses all the values in a distribution in its calculation in the sense that every value contributes to the final result in the same way that every value contributes to the calculation of the arithmetic mean. It is called the 'standard' deviation because it is the 'standard' measure of dispersion for certain types of distribution, it has practical and mathematical uses and it is standardised for all values of 'n' and can be used where this is important. It shows the dispersion of values around the arithmetic mean and in a normal curve it is known what percentage of items lie within one, two or three standard deviations on either side of the arithmetic mean.

The standard deviation can be used as a measure of dispersion in all distributions that are both symmetrical and unimodal and also in distributions that are moderately skewed. The greater the dispersion the larger the standard deviation so that a distribution with an arithmetic mean of 20 cm and a standard deviation of 12 cm has a greater or wider dispersion than a similar distribution with a standard deviation of 6 cm. In Figure 6.6 curve Z has a wider spread or dispersion than curve Y. It should be noted that *both* curves are of the 'normal curve' type and the further away a distribution is from this type the more difficult it becomes to interpret the standard deviation accurately.

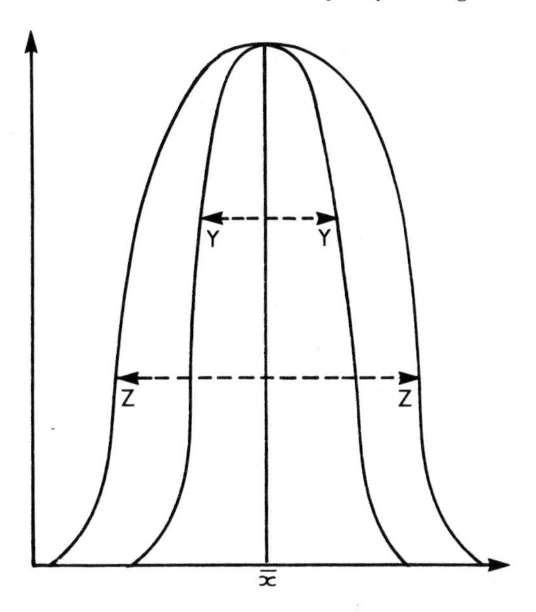

Figure 6.6 *Normal distributions*

When a normal curve has been calculated it is possible to mark off the area under the curve into certain proportions. In Figure 6.7 the arithmetic mean is shown in the centre with three standard deviations marked off on either side. It has been calculated that approximately 68 per cent of the items of a distribution will lie within one standard deviation on each side of the arithmetic mean, approximately 95 per cent of the items will lie within two standard deviations (or within 1.96 standard deviations) on each side of the mean and three standard deviations on either side of the mean will include approximately 99 per cent (or nearly all) the items in the distribution.

Although these proportions are approximations, it is possible to arrive at a fairly clear picture of a distribution if the arithmetic mean, the standard deviation and the number of items are known and it can be assumed that it is a symmetrical and unimodal distribution. If a distribution is known to have a mean of 15 cm, a standard deviation of 4 cm and 200 items, it is possible to draw the curve in Figure 6.8. The curve is constructed

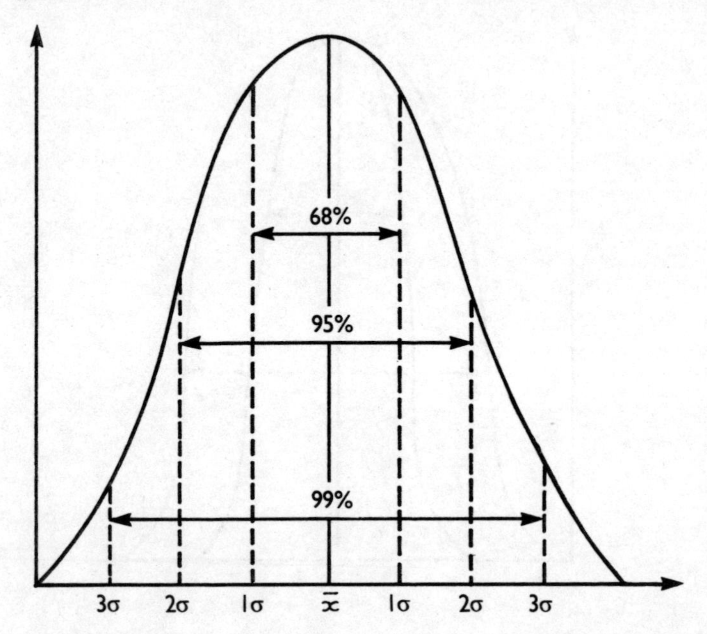

Figure 6.7 *The standard deviation and the normal curve*

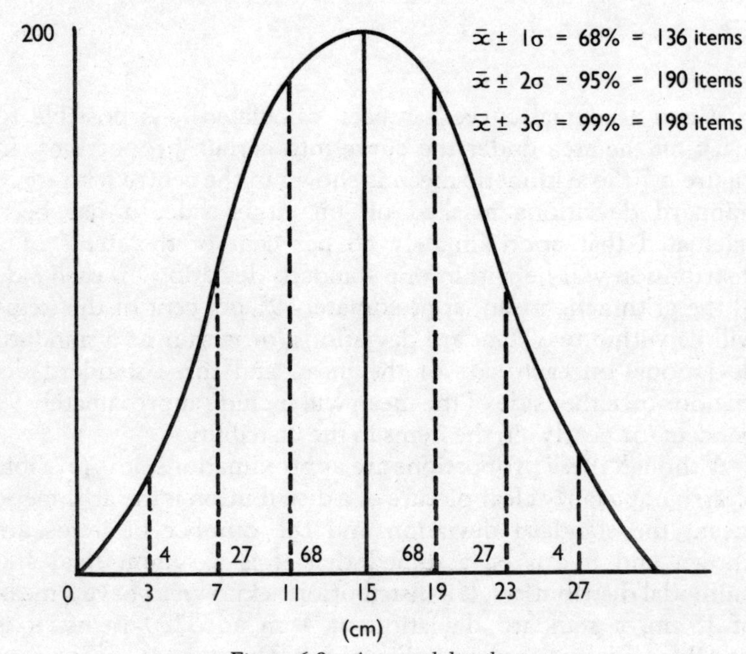

Figure 6.8 *A normal distribution*

by placing the mean in the centre of the horizontal axis, drawing in the vertical to represent the number of items and then marking the horizontal scale by adding and subtracting the standard deviation from the mean (15 cm ± 4 for one standard deviation, 15 cm ± 8 for two standard deviations and 15 cm ± 12 for three standard deviations). The shape of the curve can be arrived at approximately by assuming that 68 per cent of the items lie within one standard deviation, 95 per cent within two standard deviations and 99 per cent within three standard deviations on either side of the mean. Strictly speaking, the areas under the curve should represent these proportions.

Calculation of the standard deviation

The standard deviation is obtained by calculating the sum of the squared deviations of the individual values from the mean of the distribution, dividing this sum by the number of items in the distribution and finding the square root of the result.

In this calculation what is happening is that the deviation of each value is being taken from the arithmetic mean and then an average is being calculated from these deviations in order to control for the number of cases involved, The problem is that if the actual deviations from the arithmetic mean are added together the result will always be zero, because positive and negative deviations will cancel each other out. In order to avoid this problem, the deviations are squared, which has the effect of making the negative deviations positive; and then the square root is taken of the sum of these deviations to cancel out the squaring of them.

A simple example of this calculation using Table 6.3 should be sufficient to illustrate the methodology, because most managers will not want to make a laborious calculation when the result can be obtained from a computer program.

Scaling and adjusting

The standard deviation and the arithmetic mean can be used to adjust or scale a set of figures. For example, it may be thought that an examiner has marked a set of examination papers too severely, so that most of the candidates marked by this examiner would fail while most of those marked by other examiners would pass. The set of marks produced by the examiner can be

Table 6.3 *Standard deviation*

(1) Part-time weekly pay rates (£)	(2) Number of employees (f)	(3) Deviation of classes from class of assumed mean (d_x)	(4) (2) × (3) Frequency × deviation from assumed mean (fd_x)		(5) (3) × (4) Frequency × deviation from assumed mean, squared (fd_x^2)
0 and under 10	3	−2	−6		12
10 and under 20	11	−1	−11		11
20 and under 30	20	0	0		0
30 and under 40	12	+1		+12	12
40 and under 50	4	+2		+8	16
	50		−17	+20	51
				= +3	
	$\Sigma f = 50$	$\Sigma fd_x = 3$			$\Sigma fd_x^2 = 51$

The assumed mean is £25 (the mid-point of the class £20 and under £30):

$$\bar{x} = 25 - \frac{3}{50} \times 10 = 25 + 0.6 = 25.6$$

$$\sigma = \sqrt{\frac{\Sigma fd_x^2}{\Sigma f} - \left(\frac{\Sigma fd_x}{\Sigma f}\right)^2} \times \text{class interval}$$

$$= \sqrt{\frac{51}{50} - \left(\frac{3}{50}\right)^2} \times 10$$

$$= \sqrt{1.02 - 0.0036} \times 10$$

$$= \sqrt{1.0164} \times 10$$

$$= 1.0082 \times 10$$

$$= £10.1$$

$$\bar{x} = £25.6$$
$$\sigma = £10.1$$

68% = £25.6 ± £10.1 = approximately £15.5–£35.7
95% = £25.6 ± £20.2 = approximately £5.4–£45.8

Range: £15.5–£35.7 = 6.1 + 20.0 + 6.8 = 33 employees or 66%
£5.4–£45.8 = 1.6 + 43.0 + 2.3 = 47 employees or 94%

adjusted or scaled to correspond to the average and dispersion arrived at by the other examiners:

Examiner A:	Mean mark	40
	Standard deviation	10
Other examiners:	Mean mark	60
	Standard deviation	8

In order to find the adjusted or scaled mark corresponding to the original mark, the distance in proportion to the standard deviation is calculated.

The mark of 60 is 2 standard deviations from the mean on the original scale (40 + 10 + 10). Two standard deviations on the new scale would be 60 + 8 + 8 = 76. Hence an original mark of 30 on Examiner A's scale (ie one standard deviation from the mean) would be adjusted to a new mark of 52 (ie 60 − 8), and so on.

- *Measures of dispersion* are methods of describing the deviation or spread of a distribution.
- A *bell-shaped* or *'normal' curve* is symmetrical so that the distribution is spread equally on either side of the average.
- A *skewed distribution* has the peak of the curve on the graph displaced to left or right.
- A *Lorenz curve* is a graphical method of showing the deviation from the average of a group of data.
- The *range* is the highest value in a distribution minus the lowest value.
- The *interquartile range* is the difference between the upper quartile and the lower quartile.
- The *median* divides a distribution into half and the *quartiles* divide a distribution into quarters.
- The *standard deviation* is obtained by calculating the sum of the squared deviations of each value from the arithmetic mean, dividing by the number of values and taking the square root of the result.

  σ (for grouped frequency distribution)

  $$= \sqrt{\frac{\Sigma f d_x^2}{\Sigma f} - \left(\frac{\Sigma f d_x}{\Sigma f}\right)^2} \times \text{class interval.}$$

7
The Process of Making Statistical Decisions

Statistical decisions

One of the main distinguishing features of management is decision-making. A manager has to make choices between different courses of action and one of the functions of the use of statistics in management is to assist in making these choices. Decisions are made on emotional grounds as well as on facts. The facts provide the limits within which the manager has to choose and help to narrow the area of any disagreement that may exist.

A typical management technique, when a problem arises, is to ask for the facts. Once these have been collected the answer to the problem may be obvious. Complaints about car parking may be found to be concerned with the distance that particular people have to walk, rather than with a lack of space. A sample survey may show this very quickly, so that the answer will not be more parking space but perhaps a reorganisation of the spaces so that disabled and infirm people can park nearer their destination.

In the end, choices may have to be made on 'irrational' preferences or 'hunches'. If a company is deciding to relocate its offices, the facts may show that three sites are equally advantageous in terms of rates, rent, transport and communication facilities, distance from customers and from other company facilities, and so on. The company chairman or managing director may choose one of the three locations for a reason unrelated to the business of the organisation. He or she may have been on holiday in the chosen location, or had a relation living there once, or may prefer the sound of the name of the locality.

In the same way, the only fact a manager may need to know when choosing a new car is the price limit. The actual choice within that limit may be based on factors such as colour or the image a particular model is thought to provide.

Once choices have been made, the decisions may have to be justified, and again statistical facts may become very important.

The relocation of the company offices, for example, may be justified by emphasising the cost advantages of the new area compared with the present location, ignoring or playing down the other two possible locations.

Statistical decisions are concerned with the reliance which can be placed on figures, the confidence with which they can be used and their interpretation. They do not replace the exercise of managerial judgement based on experience or expert knowledge.

Estimation

Statistical estimation is concerned with finding a statistical measure of a population from the corresponding statistics of the sample. What this means is that when a sample survey has been carried out, the question is whether a measure such as the arithmetic mean of the sample is a good estimate of the arithmetic mean of the population from which the sample has been drawn. It is an 'estimate' because it is never certain that a sample is an exact reflection of the population itself.

This is the basis of statistical induction in that it is the process of drawing general conclusions from a study of representative cases.

Probability

Probability is the basis of sampling as described in the law of statistical regularity and the law of the inertia of large numbers. The use of probability in decision-making can be based on an analysis of past patterns of events and probabilities assigned to possible outcomes. Probability in the statistical sense can be defined as: *the probability of an event is the proportion of times the event happens out of a large number of outcomes.* An 'event' is an occurrence; spinning a coin so that it shows heads on top is an event, so is spinning a coin and not showing heads on top.

Probabilities can be regarded as relative frequencies, that is the probability that a particular event will happen. The probability of obtaining a head when a coin is spun is 50 per cent because there are two sides to the coin. If it is a 'normal' coin and has not been weighted or otherwise biased, 50 per cent of the time heads will appear on top. This will not necessarily happen if a coin is tossed only a few times but is more likely to happen the greater the number of spins.

This idea of probability can be shown by carrying out experiments. If a coin is tossed 100 times and 60 heads are thrown, then the probability of throwing a head based on this one sample can be said to be:

$$\frac{3}{5} \text{ or } 0.6.$$

$$\text{Probability} = \frac{\text{the total number of ways an event can happen}}{\text{the total number of outcomes to the experiment}}$$

There is only one way that heads can be spun, and there are two possible outcomes to the experiment, heads or tails. The probability of throwing a head is:

$$\frac{1}{2} \text{ or } 0.5.$$

The larger the number of throws the more reliable the information becomes in the sense that it becomes increasingly likely that the 50/50 balance will happen. Each set of throws is a sample of an infinite number of throws and the bigger the sample the more reliable it will be. This is an important factor in deciding on the size of sample surveys.

When an event is certain to happen, the probability of its happening is:

$$1 \text{ or } \frac{1}{1}$$

because there is only one way the event can happen and there can be only one outcome to the experiment. When an event can never happen, the 'probability of its happening' is 0! This means that all probabilities have a value between 0 and 1.

The 'sample space' is the total number of possible outcomes. When a die (or dice) is rolled there are six possible outcomes (the sample space (S) = 1, 2, 3, 4, 5, 6); the number of ways any one number can be uppermost is one, so that:

$$P = \frac{1}{6}.$$

The number of ways uneven numbers may appear on top is three (1, 3, 5):

$$P = \frac{3}{6} = \frac{1}{2}.$$

When two dice are thrown at the same time and the different possible combinations are noted, the sample space will be 21 (ie there are 21 possible combinations although only 11 different scores). The probability of throwing a double 3 is:

$$P = \frac{1}{21}.$$

The probability of throwing a score of 6 with the dice is:

$$P = \frac{3}{21} \text{ or } \frac{1}{7}$$

because there are three possible ways the event can happen $(1+5, 2+4, 3+3)$.

When deciding on the sample space, as in other matters, definition is important. When the dice are thrown at the same time and it is the different scores which are noted, the sample space will be 21, as shown by the crosses in Figure 7.1.

If, however, it is important which die shows which number in

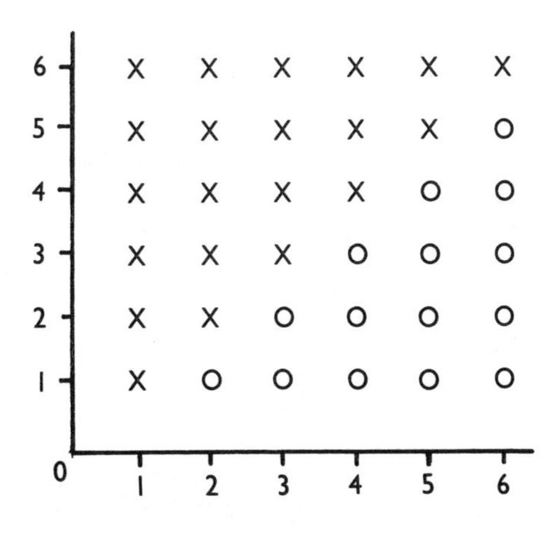

Figure 7.1 *Sample space*

a combination, the sample space rises to 36, as shown by the crosses and circles in Figure 7.1. If one die is red and one green, there are two possible combinations to arrive at a total throw of 3: *R*2 and *G*1 or *R*1 and *G*2.

As well as 'sample space' diagrams such as Figure 7.1, tree diagrams are used to reduce the need to produce a complete sample space for complicated combinations. For example: 'what is the probability of drawing two kings from a pack of 52 cards (replacing the first card before drawing the second)?' The probability of the first card drawn being a king is:

$$\frac{1}{13} \text{ ie } \frac{4}{52}.$$

The probability of not drawing a king is:

$$\frac{12}{13}.$$

The replacement of the first card means that the probabilities remain the same for the second chance. If this process were repeated a great number of times:

$$\frac{1}{13}$$

of the draws would produce a king on the first draw and out of these draws:

$$\frac{1}{13}$$

would produce a second king. This means that:

$$\frac{1}{13} \text{ of } \frac{1}{13}$$

of the draws would be expected to produce two kings:

$$\text{The probability of drawing two kings } (P) = \frac{1}{13} \times \frac{1}{13} = \frac{1}{169}.$$

These proportions can be shown as a tree diagram (Figure 7.2).

Probability is the basis for **statistical tests** which can be used when it is possible to hypothesise a certain probability of success, when trials are independent of each other and when the number of trials is relatively small. Statistical tests involve

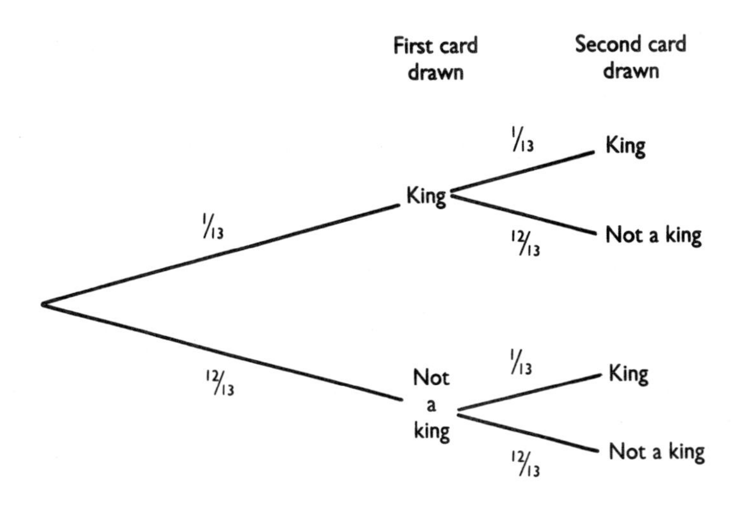

Figure 7.2 *Tree diagram*

probability distributions which are similar to frequency distributions, except that probabilities are used instead of frequencies.

A **binomial distribution** is concerned with two items, such as the probability that an event occurs and the probability that it does not occur.

The **poisson distribution** provides a good approximation to the binomial distribution. Both distributions are discrete, while the normal distribution is continuous, and they can be used whenever there are events which occur in a random manner.

The **sign test** for example, is used in simple 'before and after' experiments when a small number of cases are used to determine whether the experiment has been successful or not.

If the experiment is concerned, for example, with whether or not a promotional campaign increases sales in a company's outlets, a sign is designated for successes (say +) and for failures (say −). Then, making assumptions about both the population and the method of sampling, the level of probabilities of outcomes can be assessed.

Probability and management

From a manager's point of view, the most useful aspect of probability is how often an event will normally occur compared with how often it could occur.

If it is reported of a company that it delivers its goods on time 70 per cent of the time, or:

$$\frac{7}{10}$$

then this means that records show that 7 times out of 10 the goods are delivered on time, and this can be set against the fact that they could deliver their goods on time 10 times out of 10. The managers receiving the goods can take the 70 per cent delivery record as a factor in their planning.

Electricity and gas corporations who collect the payment of their bills on a regular quarterly basis, can predict how many bills will be paid by the due date and how many after the first or second reminder, by looking at their records.

Their records will show which customers pay their bills immediately every time, those who need a reminder on occasions and those who never pay until they have received a reminder. The cash flow can be predicted quite accurately from such records. At the same time, the level of prompt payment can be improved by introducing such methods as direct debiting and budget account systems which do not rely on reminders.

In the management planning process, as well as comparing how often an event will normally occur as against how often it could occur, it is useful also to remember that events occur in different ways:

i *Mutually exclusive events*: there are two possible outcomes from spinning a coin, a head or a tail. These events are mutually exclusive because only one can happen, the occurrence of one of them excludes the occurrence of the other.

ii *Independent events*: if a coin is spun on two different occasions and 'heads' is uppermost each time, the two 'events' are independent, because the second spin of the coin could not be influenced by the first spin.

iii *Conditional events*: in order to obtain two 'heads' a coin has to be spun at least twice and the result of the first spin must be 'heads' for there to be any probability of the second spin producing the desired result. The second event is subject to the first event having taken place.

In management terms it is useful in decision making to know whether events are mutually exclusive, independent or conditional. If an organisation is planning to sell its town centre site and move to a 'green-belt' site, choosing one site may exclude the choice of others. At the same time, the move may be dependent on the sale of the present site. These may be important factors in the production of a flow chart to assist in the planning process and to show the actions required at each stage.

A simple flow chart may show the following 'solutions' to the organisation's 'problem':

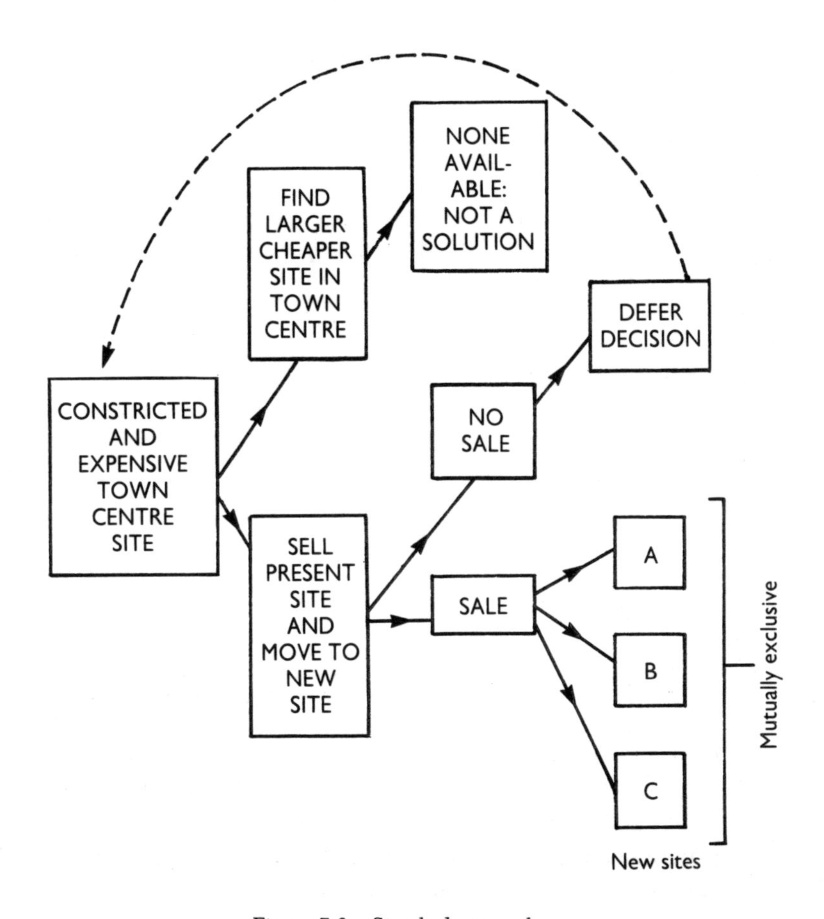

Figure 7.3 *Simple decision chart*

Sampling and probability

Standard error

If a sample is selected at random from a population, there is a high probability that it will represent the population from which it is drawn and therefore provide evidence for management decision-making. If a measure, such as the arithmetic mean, is looked at from each sample of a large number of samples taken from a population, most of these sample means will be the same or very similar, although some may by chance be above or below the other means. When all these means are placed on a graph this is likely to appear as a normal curve with most sample means in the centre and a few on either side (Figure 7.4).

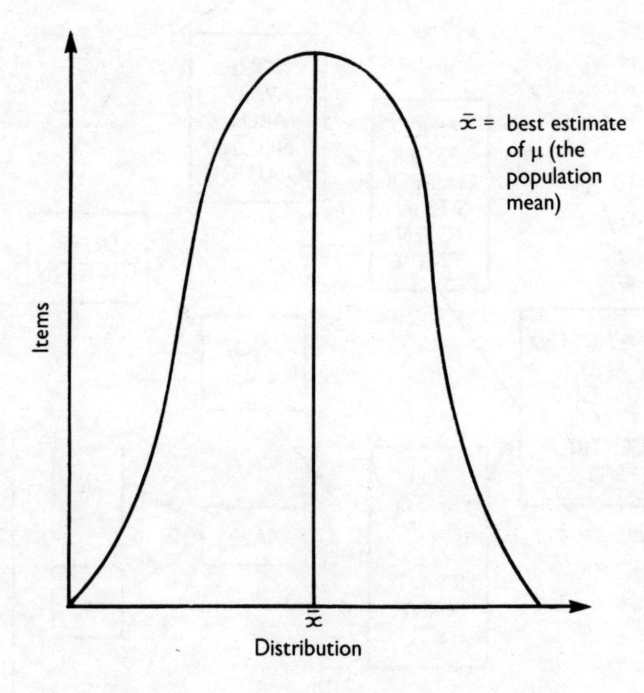

Figure 7.4 *The sampling distribution of the mean*

The average of these means will be the 'best estimate' of the population mean. The standard deviation of this distribution is called the standard error and this is found by relating the sample size to this standard deviation:

$$\text{Standard Error (SE)} = \frac{\text{standard deviation of the sample}}{\sqrt{\text{sample size}}}$$

$$= \frac{\sigma}{\sqrt{n}}$$

where n = sample size

σ = standard deviation of the sample

In Chapter 4, on collecting management information and on probability and sampling, it was pointed out that the accuracy of a sample was not dependent on the population size. A population of five million does not require a larger sample than a population of 50,000. The accuracy of a sample depends on its own size and on the variability of the characteristics measured. If the 'population' is homogeneous, such as items from a production line, the sample can be comparatively small. If it is not homogeneous and contains many variations, such as a population's political or social opinions, then the sample needs to be sufficiently large to include a good representation of these opinions.

A number of samples taken from a production line could be expected to have very similar results for such measures as the arithmetic mean and dispersion because each unit is meant to have the same measurements as other units. A number of samples taken from the general public as part of an opinion survey, on the other hand, could be expected to produce variations in the arithmetic mean for each sample and a relatively wide dispersion of these around the population mean. In order to improve the accuracy of a sample survey, the size of the sample can be increased.

For example, a sample of 500 people may have a standard deviation of 10. The standard error will be:

$$SE = \frac{\sigma}{\sqrt{n}} = \frac{10}{\sqrt{500}} = \frac{10}{22.36} = 0.45.$$

If the sample size is increased to 1,000 the standard error will be:

$$SE = \frac{10}{\sqrt{1{,}000}} = \frac{10}{31.62} = 0.32.$$

In fact, in order to halve the original standard error, it is necessary for the sample size to be increased four times:

$$SE = \frac{10}{\sqrt{2{,}000}} = \frac{10}{44.72} = 0.22.$$

This is an important consideration for managers commissioning sample surveys, because the increase in costs and time may not be worth the decrease in standard error.

Although the frequency distribution of a population may not correspond very closely to the normal curve, the sampling distribution will, providing the sample size is sufficiently large (say over 30 units or people). This enables the accuracy of samples taken from any population to be estimated. In considering the interpretation of the standard deviation in the section on dispersion (see Chapter 6) it was noted that 95 per cent of the items in a normal distribution lie within two standard deviations of the mean of the distribution.

Confidence

In terms of standard error it follows that 95 per cent of all the means of all the samples must lie within two standard errors of the true mean of the population since the distribution of the means is normal with a mean equal to the population mean. If a single sample is taken, 19 times out of 20 the sample mean will be within two standard errors of the true mean of the population. When a manager is considering the results of a sample survey and is told that the arithmetic mean is, say, 15 with a standard error of 0.25, the manager can be *confident* that 19 times out of 20 the population mean will be between 14.5 and 15.5 ($15 \pm 2 \times 0.25$). If this is not a sufficiently accurate estimate for the purposes required of the result, then the sample size can be increased to reduce the standard error. If the sample size was increased four times and the same arithmetic mean was found, the standard error would be 0.125 and the population mean would lie between 14.75 and 15.25.

Significance

If a manager believes the population mean to be 19 this can be tested by arranging a sample survey. This may be based on a sample of 200 units and produce a result showing a mean of 18 and a standard deviation of 5. The standard error will be 0.35. Two standard errors on either side of the assumed population mean (19) will give values of 18.3 and 19.7. The sample mean (18) is outside the range of the values and therefore the difference between the assumed mean and the sample mean can be said to be *significant*. It is significant because there is evidence to suggest, from the sample result, that the population mean is not 19.

If the sample result had shown an arithmetic mean of 18.5 this would have meant that the difference between the two means was *not significant* and the population mean could be 19.

Quality

One management application of these statistical techniques is the control of quality. This can be described as statistical process control when the emphasis is on the process of production rather than on the product.

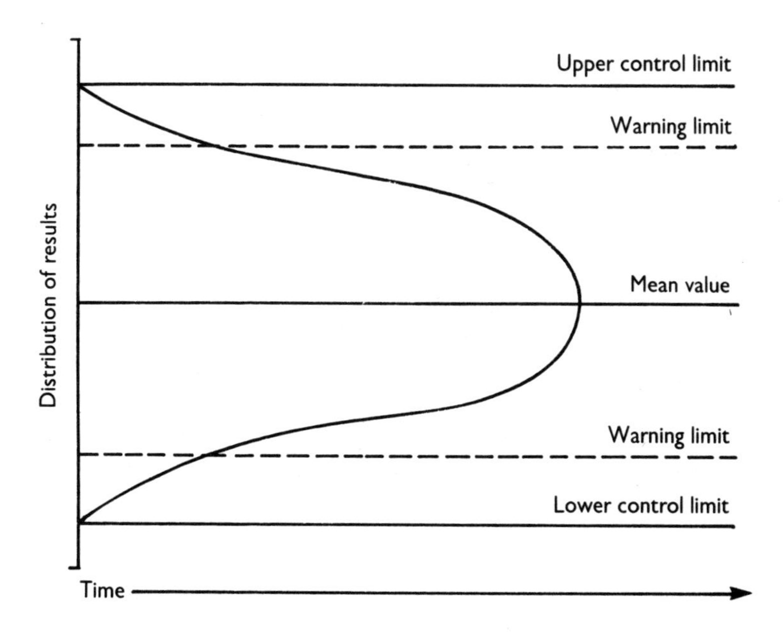

Figure 7.5 *Quality control*

The technique is to establish a system of measurement for a production line so that samples of the product can be observed as the units are produced. The results of these samples can be placed on to a control chart (see Figure 7.5) in order to observe the performance of the production process. The control chart will have upper and lower limits which help to establish boundaries of 'tolerance'. If the results show that the process is moving outside these boundaries this will indicate that there has been a change in the production process which needs investigation.

The sample results are likely to be similar to a normal curve. This means that approximately 68 per cent of the results should be within one standard deviation on either side of the arithmetic mean, approximately 95 per cent within two standard deviations and approximately 99 per cent within three standard deviations. The upper and lower control limits can be set at the 99 per cent level with a warning limit at 95 per cent, so that if there is a variation in the results moving outside the range where 95 per cent of the population have been shown to lie, it can be

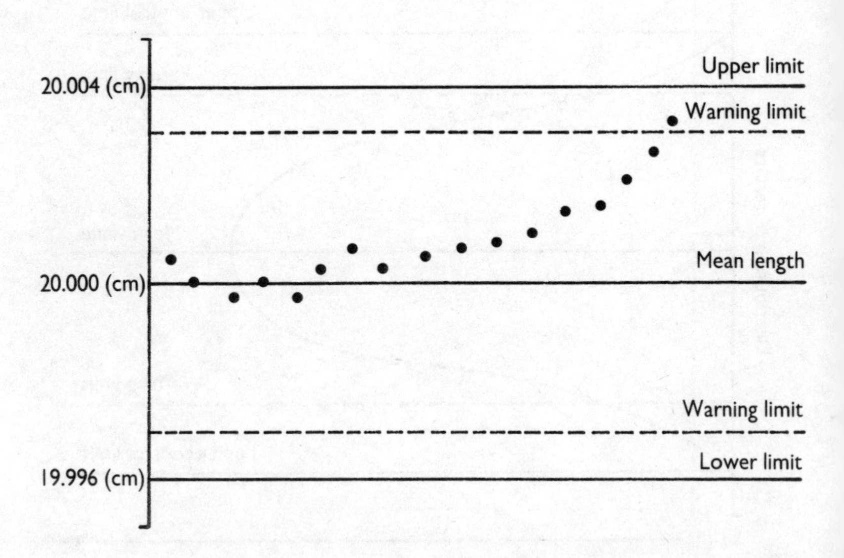

Figure 7.6 *Control chart*

deduced that there may be a problem. A process control chart can be plotted to illustrate this (see Figure 7.6).

Correlation

Correlation is concerned with whether or not there is any association between two variables. This is an important feature of decision-making, because knowledge of relationships between variables, based on evidence, provides control over events and enables plans to be formulated. For example, if a manager plots on a graph the level of expenditure on marketing against the level of sales revenue, the variables may show a strong correlation.

In Figure 7.7, as more is spent on marketing, so sales revenue increases and there is what is termed 'strong, positive correlation'.

This 'scatter diagram' shows that there is a pattern between the points so that a 'line of best fit' can be drawn. This is a

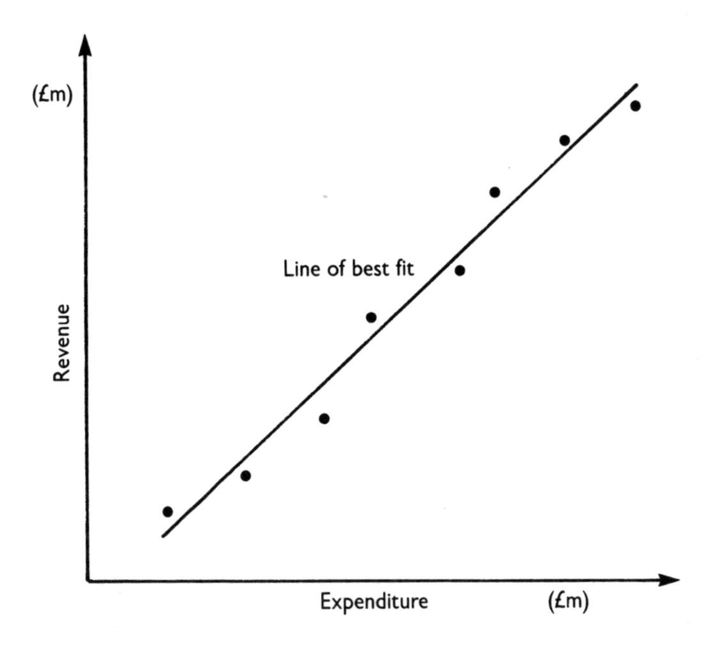

Figure 7.7 *Correlation scatter diagram*

straight line which best fits the pattern made by the points, coming as close to as many of them as possible. In this case the correlation shown is linear (it approximates to a straight line). Negative linear correlation would arise if, as expenditure rose, revenue fell. Where the points are all over the place it can be said that there is no correlation.

The results shown in Figure 7.7 do not 'prove' that an increase in marketing expenditure will increase sales revenue. Although the graph indicates that there appears to be an association between the two variables, it does not prove any causal relationship. Sales revenue may be rising 'in spite of' increased marketing expenditure. The marketing plan could be very poor and at the same time sales could be boosted by a competitor going out of business.

Correlation is a starting point for an investigation and, as usual, the interpretation of the results needs to be made with great care and using all the experience and judgement of a manager. There may appear to be negative correlation because of a time lag. The effect of a promotion organised by the marketing department may take time to be noticed in an increase in sales. If a product or service is at a stage of decline in its life cycle, sales may be falling, while extra expenditure is being used to prepare a promotional campaign to revive sales.

Correlation coefficients

There are two commonly used coefficients to indicate correlation:

(a) the 'product moment' coefficient which measures linear correlation

(b) the rank correlation coefficient which measures correlation 'ordered' or 'ranked' on the interval scale.

These coefficients can show the strength of the association between two variables, the direction of this association (positive or negative), and the variability in the relationship.

The correlation coefficient is '*r*' for a product moment correlation and '*r*'' for a rank correlation and the coefficient will always be between -1 and $+1$. Strong positive correlation is indicated by $+0.9$, while -0.9 indicates strong negative correlation. The figures $+0.1$ or -0.1 indicate weak positive and negative correlation and even ± 0.6 or ± 0.7 do not show very strong correlation.

If the correlation coefficient is ± 0.8 or ± 0.9, the association between the two variables can be said to be strong. If r is 0.9, r^2 will be 0.81 and it can be said that 81 per cent of the variability in one variable can be accounted for by its linear relationship with the other variable. It can be seen from this that if $r = 0.8$, 64 per cent of the variability in one variable can be accounted for by its linear relationship with the other variable and that if $r = 0.7$ the figure is only 49 per cent.

Calculation of the product moment coefficient of correlation

This is based on a formula which divides the mean product of the deviations from the mean, by the product of the standard deviations:

$$r = \frac{\Sigma(xy)}{\sqrt{\Sigma x^2 \Sigma y^2}}$$

where:

- r is the product moment coefficient
- x is the deviation from the mean of one variable
- y is the deviation from the mean of the other variable
- $\Sigma(xy)$ represents the sum of the products of the deviations from the mean.

Table 7.1 *Product moment correlation coefficient*

Year	*Marketing (£10,000)*	*Sales (£100,000)*	*x* $(\bar{x} = 8)$	*y* $(\bar{y} = 12)$	x^2	y^2	xy
1990	4	10	−4	−2	16	4	8
1991	6	12	−2	0	4	0	0
1992	10	11	+2	−1	4	1	−2
1993	12	15	+4	+3	16	9	12
	32	48			40	14	18

$$\text{(Marketing) } \bar{x} = \frac{32}{4} = 8 \qquad \Sigma(xy) = 18, \Sigma x^2 = 40, \Sigma y^2 = 14$$

$$\text{(Sales) } \bar{y} = \frac{48}{4} = 12$$

$$r = \frac{18}{\sqrt{40 \times 14}} = \frac{18}{23.66} = +0.76$$

The coefficient shows a positive correlation between the two variables.

The calculation of the rank coefficient of correlation

This is based on a formula which squares the differences between the rank of one variable against the rank of the other variable:

$$r' = 1 - \frac{6\Sigma d^2}{n(n^2 - 1)}$$

where:

r' is the rank correlation
Σd^2 is the sum of the squares of the differences between the ranking
n is the number of items.

For example, in a marketing survey on colour matching, a panel of experts is asked to judge the colour shades of ten paints and to arrive at a ranking of 1–10, the darker the shade,

Table 7.2 *Rank correlation coefficient*

Paints	(1) *Panel of experts*	(2) *Panel of consumers*	(3) *d* *(2) – (1)*	(4) d^2 $(3)^2$
A	10	10	0	0
B	9	6	3	9
C	8	7	1	1
D	7	5	2	4
E	6	9	3	9
F	5	5	0	0
G	4	4	0	0
H	3	2	1	1
I	2	3	1	1
J	1	1	0	0
				$\Sigma d^2 = 25$

$$r' = 1 - \frac{6 \times 25}{10(100-1)} = 1 - \frac{150}{990}$$

$$= 1 - 0.15 \qquad = 0.85$$

the higher the rank. Another panel, chosen at random from consumers, is asked to carry out the same process of judging and ranking the ten shades. The results are shown in Table 7.2.

This shows a high level of positive correlation between the views of the two panels.

This result is one element in the marketing research process and other tests may be carried out in order to check on this result. Statistical decision-making is not a process for arriving at foolproof answers. Managers will be only too aware that such a process does not exist. It does involve methods and techniques which can help in decision-making, planning and control.

- *Statistical decisions* are concerned with the reliance which can be placed on figures, the confidence with which they can be used and their interpretation.
- *Statistical estimation* is concerned with finding a statistical measure of a population from the corresponding statistic of the sample.
- *Statistical probability* is the proportion of times an event happens out of a large number of outcomes:

$$\text{Probability} = \frac{\text{the total number of ways an event can happen}}{\text{the total number of outcomes to the experiment}}$$

- *Standard error* $= \dfrac{\sigma}{\sqrt{n}}$
- Correlation is concerned with whether or not there is any association between two variables.
- *Product moment correlation* $= \dfrac{\Sigma(xy)}{\sqrt{\Sigma x^2 \Sigma y^2}}$
- *Rank correlation coefficient* $= 1 - \dfrac{6\Sigma d^2}{n(n^2-1)}$

8
Forecasting for Managers

Planning and control in management

Managers need to forecast trends whether they are in the private or public sectors or in a non-profit organisation. Forecasting is an important element of planning and control and it is, therefore, an essential element of management which becomes more important as a manager's responsibility increases. At an operational level a manager will be more concerned with day-to-day problems as they arise than with forecasting and planning in the medium or long term. It is, however, very important in all types of organisation that trends are charted at an appropriate level of management so that forward planning can be carried out.

Management decisions are often based on an idea of what is likely to happen in the future. Whether to take on extra staff or leave vacancies unfilled will be a decision based on the state of the order books and judgement of whether the work of the organisation is going to rise or fall. This is just as true in an organisation such as a charity as in a commercial organisation such as an insurance company. Investment decisions are equally based on judgement of future prospects and marketing decisions are based on the research into trends within an organisation's markets.

In arriving at a sales forecast, for example, a company may start with an economic forecast, which will consider trends within the whole economy. This may be based on indicators such as the level of unemployment, trends in interest rates and in inflation and in the balance of payments position. High interest rates will make investment expensive and reduce consumers' disposable income. This information can be translated into an industry forecast, so that if an organisation is in housing the market may be judged to be likely to fall fairly quickly in response to high interest rates, while the food market may remain fairly stable in the same conditions. From the industry forecast a manager can produce a company sales forecast depending on the share of the market and the ability of the company to compete in the prevalent market conditions.

The company sales forecast will depend also on such factors as surveys of buyers' intentions, on the salesforce decisions and expert views. Statistical techniques can be used with averages, measures of dispersion and correlation providing a starting point. Index numbers and time series analysis are both statistical techniques used to assist this process of planning and forecasting. In line with other statistical techniques, they can provide the factual evidence on which managers are able to base their judgements.

Index numbers

What is an index number?

An index number is a measure designed to show average changes in a variable, such as price or quantity, of a group of items over a period of time. The objective of producing an index number is to provide a measure which simplifies comparison over time, by replacing complicated figures with relatively simple ones calculated on a percentage basis.

Most indexes (or indices) are weighted averages and therefore they provide only a part of the whole picture, an indication rather than a complete description. The index of retail prices, for example, may provide a good description of changes in the inflation rate for the 'average' household, but a less good description for the many households that are not 'average'. It may be, however, the best general indicator available.

Calculation of an index number

A price or value index is found by the formula:

$$\text{Price index} = \frac{p_1}{p_0} \times 100$$

where p_0 is the price in the base year and p_1 the price in the year to be compared with it.

For example: if the price of bread is £1 in 1990 and £1.50 in 1995 the index would be:

$$\text{Price index} = \frac{1.50}{1.00} \times 100$$

$$= 150.$$

This in fact means that the price has risen by 50 per cent over the five years, assuming that the base year is equal to 100.

When more than one commodity is included in an index, account has to be taken of the relative importance of each commodity to the total expenditure.

For example: if the price of milk rose from £0.30 in 1990 to £0.60 in 1995 the combined index could be calculated as follows:

$$\frac{150+200}{2} = 175.$$

It is possible, however, that milk is only half as important as bread in the 'average' household so that to say that prices had risen by 75 per cent would not be very accurate.

If it is assumed that the total expenditure of a household is on bread and milk, it may be that spending on bread is two-thirds of the total while spending on milk is one-third. A rise in the price of bread would have a greater impact on the household than an equivalent rise in the price of milk.

The combined index could be calculated as shown in Table 8.1.

Table 8.1 *Calculating a combined index*

	(1)	(2)	(3)	(4)	(5)
	1990 price (£)	*1995 price (£)*	*Increase in price %*	*Weight*	*Product (3) × (4)*
Bread	1.00	1.50	50	2	100
Milk	0.30	0.60	100	1	100
				3	200

$$\text{Price Index} = \frac{200}{3} = 66.66\% \text{ or } 166.66$$

This would be a weighted index and might be felt to represent a better indication of price changes than an unweighted index.

The UK Index of Retail Prices (IRP) is based on this assumption. The Family Expenditure Survey is designed to provide information on the 'average' household's 'typical' basket of goods and services and the proportion they each represent of total household expenditure. Weights are then revised annually to show changing comparisons between such items as food, housing and transport. This forms the best indicator available of inflation and changes in the cost of living. Any particular household will tend to have a slightly different rate of inflation, depending on how closely their weighted 'basket of goods' compares to that of the Index.

Managers need to be aware that index numbers are based on the 'average' and are usually 'weighted'. They need to look also at the choice of the base year for an index because this can influence the rise and fall in prices, costs and values. If the base year chosen is during a 'slump', the prices and costs are likely to be depressed as compared with the following 'boom' years. At the least, the interpretation of index numbers needs to be considered with these factors in mind.

If the news is that inflation is rising at 6 per cent a year on a month-by-month basis, this means that the IRP has risen by six points from a base of 100 (say from 142 to 148) in the last 12 months (say June 1990 to May 1991 inclusive). The next month's figure will be based on July 1990 to June 1991 and will reflect changes in average prices during these particular months.

Wage negotiations are often based on 'keeping up with the inflation rate' or on 'not fuelling the rate of inflation' by causing price rises in order to cover wage rises. The Index of Industrial Production is used to show changes in the level of industrial output and indices can be used for the value of imports and exports, the cost of raw materials and fuel and power, changes in unemployment, house prices and so on. In general, managers are able to use index numbers as indicators in planning future costs and expenditure levels and in assessing trends in the economy, levels of employment, levels of production and consumer expenditure.

Time series

What is a time series?

A time series consists of numerical data recorded at intervals of time. In this sense index numbers are a time series. Graphs are the most popular forms of presentation because they clearly illustrate the relationship between the other variables and time in showing the trend.

Calculation of a moving average

There are a range of calculations designed to produce a trend line, some of these are based on trends in order to produce a straight 'line of best fit'. For most purposes a line drawn by eye or using averages over periods of time can be used to smooth out fluctuations and show the general trend. Regression analysis is used as a more advanced technique for similar purposes.

The most commonly used non-linear trend is the moving average. This is a process of repeatedly calculating a series of different average values along a time series in order to produce a trend line.

For example: if a company's revenue is tabled over nine years, the following results may arise:

Table 8.2 *Moving average*

Years	*Revenue (£m)*	*3-year moving average*	*5-year moving average*
1984	24		
1985	20	19.67	
1986	15	19	22.6
1987	22	23	23
1988	32	26.67	23.8
1989	26	27.33	27.8
1990	24	28.33	31.4
1991	35	33	
1992	40		

The three-year moving average is calculated by adding the revenue for the first three years and dividing by three ($59 \div 3 \simeq 19.67$). The first year is subtracted and the next year added in order to find the next moving average:

$$(-24+22 = -2+59 = 57 \div 3 = 19)$$

and so on. The five-year moving average is found by the same method:

$$(24+20+15+22+32 = 113 \div 5 = 22.6).$$

The results can be charted on a graph (see Figure 8.1)

In Figure 8.1 the actual revenue is charted over the nine years.

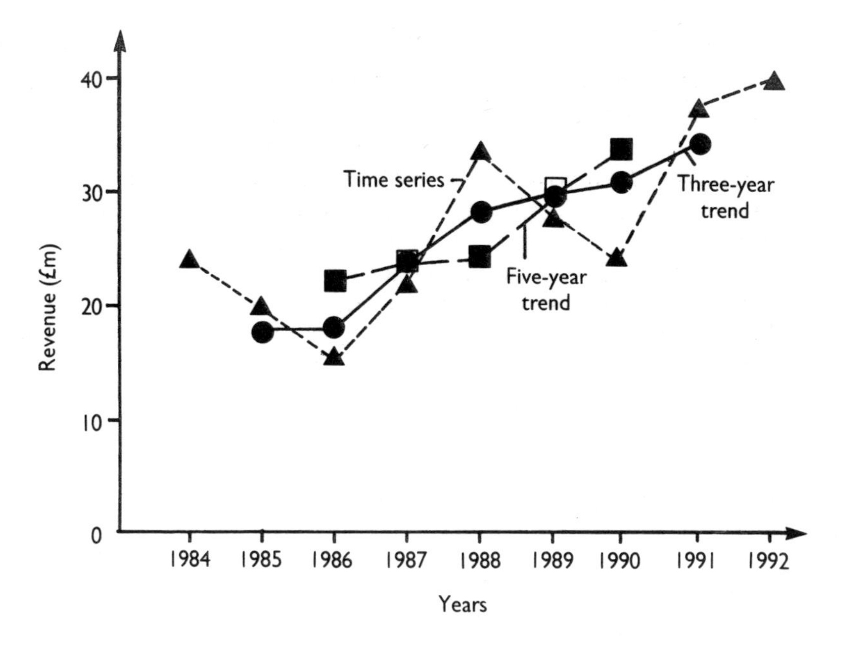

Figure 8.1 *Moving average*

The three-year moving average flattens out the fluctuations and shows the general upward trend in revenue. The five-year moving average smooths out the fluctuations even further and shows a clear upward movement. This is in fact close to a straight line of 'best fit'.

The graph emphasises the trend shown in the figures. Peaks of revenue occur every few years with the troughs occurring at similar intervals. The overall trend is upwards, although further information would be needed in order to judge whether this is due to inflation or to a change in real income. A manager could use such graphs to forecast trends over the next few years. These trends may not continue, of course, but assuming that they did, a manager could hope for a generally rising trend of revenue with a fall over the next two or three years to perhaps £30m followed by a rise to a new high peak. All of this would depend on 'other things remaining equal', so that there are not dramatic changes in competition, government policy and other factors on the horizon.

Seasonal variations

Many products and services are affected by seasonal variations in demand and this in turn affects an organisation's revenue and profit. These variations are often eliminated from data in order to identify the underlying trends. Tables are often produced with a footnote to the effect that the figures are 'seasonally adjusted'.

For example, unemployment is very prone to seasonal variations, with high levels in the winter and much lower levels in the summer. These fluctuations occur alongside any more general trend of rising or falling rates of unemployment over a period of

Table 8.3 *Seasonal adjustment*

Year and quarters		(1) *Revenue (£m)*	(2) *4-quarterly total*	(3) *Centred totals*	(4) *Trend ([3]÷8)*	(5) *Variations from the trend (1)–(4)*
1990	1	24				
	2	15				
			75			
	3	16		160	20	−4
			85			
	4	20		181	22.63	−2.63
			96			
1991	1	34		198	24.75	+9.25
			102			
	2	26		214	26.75	−0.75
			112			
	3	22		232	29.00	−7.00
			120			
	4	30		252	31.5	−1.5
			132			
1992	1	42		270	33.75	+8.25
			138			
	2	38		278	34.75	+3.25
			140			
	3	28				
	4	32				

Year	*Quarters* 1	2	3	4
1990	—	—	−4	−2.63
1991	+9.25	−0.75	−7	−1.5
1992	+8.25	+3.25	—	—
	+17.50	+2.5	−11	−4.13
Average	+8.75	+1.25	−5.5	−2.07
Adjustment	+0.61	+0.61	+0.61	+0.61
Seasonal variation	+9.36	+1.86	−4.89	−1.46

Adjustment = 8.75 + 1.25 − 5.5 − 2.07 = 2.43 ÷ 4 = +0.61

years. There are a number of methods of adjusting for seasonal variations. The example in Table 8.3 shows a frequently used method which illustrates the general principles involved.

Table 8.3 shows the revenue of an organisation over three years divided into quarters (or seasons). The four quarterly totals are calculated:

$$(24+15+16+20 = 75, 15+16+20+34 = 85 \text{ and so on}).$$

These are then 'centred' by adding together each pair of four quarterly totals:

$$(75+85 = 160, 85+96 = 181 \text{ and so on})$$

and these are divided by 8:

$$(160 \div 8 = 20, 181 \div 8 = 22.63 \text{ and so on}).$$

Column (5) shows the variation between the actual revenue and the trend (Column (1)–Column (4)). Column (5) is then used to calculate the seasonal variations as set out in Table 8.3.

In Table 8.3 the average of each quarter is found by dividing the total for each quarter by the appropriate number of quarters, in this case 2:

$$(+17.50 \div 2 = +8.75, +2.50 \div 2 = 1.25 \text{ and so on}).$$

These averages are added together to provide an overall 'average' and this is divided by 4 to arrive at an 'adjustment' figure

Table 8.4 *Seasonally adjusted figures*

Year and quarters		*Revenue (£m) seasonally adjusted*	*Adjusted for revenue (to two significant figures)*
1990	1	24 − 9.36 = 14.64	15
	2	15 − 1.86 = 13.14	13
	3	16 + 4.89 = 20.89	21
	4	20 + 1.46 = 21.46	21
1991	1	34 − 9.36 = 24.64	25
	2	26 − 1.86 = 24.14	24
	3	22 + 4.89 = 26.89	27
	4	30 + 1.46 = 31.46	31
1992	1	42 − 9.36 = 32.64	33
	2	38 − 1.86 = 36.14	36
	3	28 + 4.89 = 32.89	33
	4	32 + 1.46 = 33.46	33

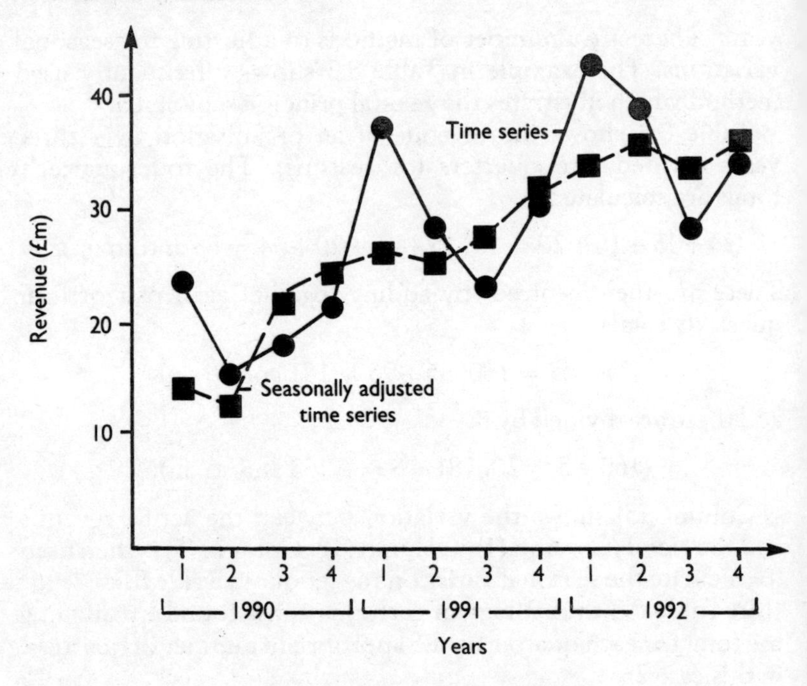

Figure 8.2 *Seasonally adjusted series*

($+2.43 \div 4 = +0.61$). This is then added to each quarter equally to provide a seasonal variation adjustment. These can then be used to produce a seasonally adjusted time series (see Table 8.4).

The original revenue figures and the adjusted figures can then be compared on a graph, as in Figure 8.2.

It should be noted in Table 8.4. that the seasonal adjustment minus figures have been added and the plus figures subtracted. This is because additions are made to the generally low seasons and subtractions are made from the high seasons in order to reduce seasonal variations. This has the effect of reducing the peaks and troughs in order to show an 'underlying' trend more clearly. The general trend of the seasonally adjusted trend line indicates that revenue is rising although it could be judged that there are signs that a 'plateau' may have been reached with revenue levelling out.

Irregular fluctuations

A time series may be affected by irregular or residual factors.

Table 8.5 *Irregular factors*

Year and quarters		*Original series*	=	*Trend*	±	*Seasonal variation*	±	*Irregular factors*
1990	1	24						
	2	15						
	3	16		20		−4.89		+0.89
	4	20		22.63		−1.46		−1.17
1991	1	34		24.75		+9.36		−0.11
	2	26		26.75		+1.86		−2.61
	3	22		29.00		−4.89		−2.11
	4	30		31.5		−1.46		−0.04
1992	1	42		33.75		+9.36		−1.11
	2	38		34.75		+1.86		+1.39
	3	28						
	4	32						

These are external factors, and if they are large the forecasts are likely to be less reliable than if they are small. Once the trends and seasonal adjustments have been calculated, the value of the irregular factors can be found. The irregular or residual factors are added to or subtracted from the seasonal variations and the trend to arrive back at the original series.

Original Series = Trend ± Seasonal Variations ± Irregular Factors

This is shown in Table 8.5.

Linear trends

If there is a linear relationship between variables, trends can be shown by a straight line. Regression analysis and calculations by the least squares method are advanced techniques for arriving at a line of best fit. A simple approach, which illustrates the linear trend, is to calculate a number of averages for various parts of the data. For example, three averages can be calculated from the data in Table 8.2.

A linear trend line such as that shown in Figure 8.3 can give an impression of steady growth while the moving averages shown in Figure 8.1 indicated a more complex situation. This straight line does show the basic trend and this can be useful when it is used with the actual time series.

Table 8.6 *Linear trend*

Years	*Revenue (£m)*	
1984	24	
1985	20	
1986	15	Arithmetic mean of years 1984–87 = $\frac{81}{4}$ = 20.25
1987	22	
1988	32	Arithmetic mean of total sales = $\frac{238}{9}$ = 26.44
1989	26	
1990	24	Arithmetic mean of years 1989–92 = $\frac{125}{4}$ = 31.25
1991	35	
1992	40	

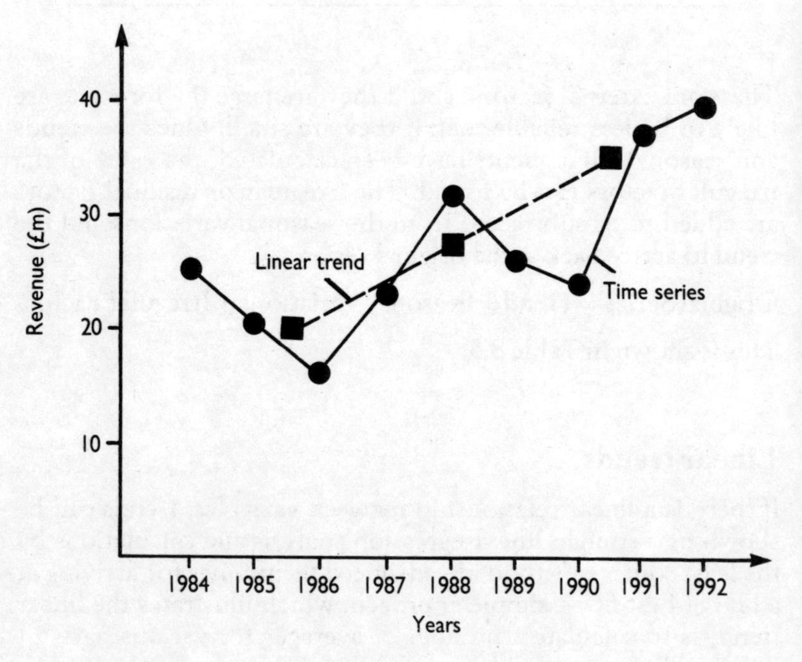

Figure 8.3 *Linear trend line*

The Z chart

A graphical method used to illustrate the trend against the original data is the Z chart. This consists of three graphs plotted

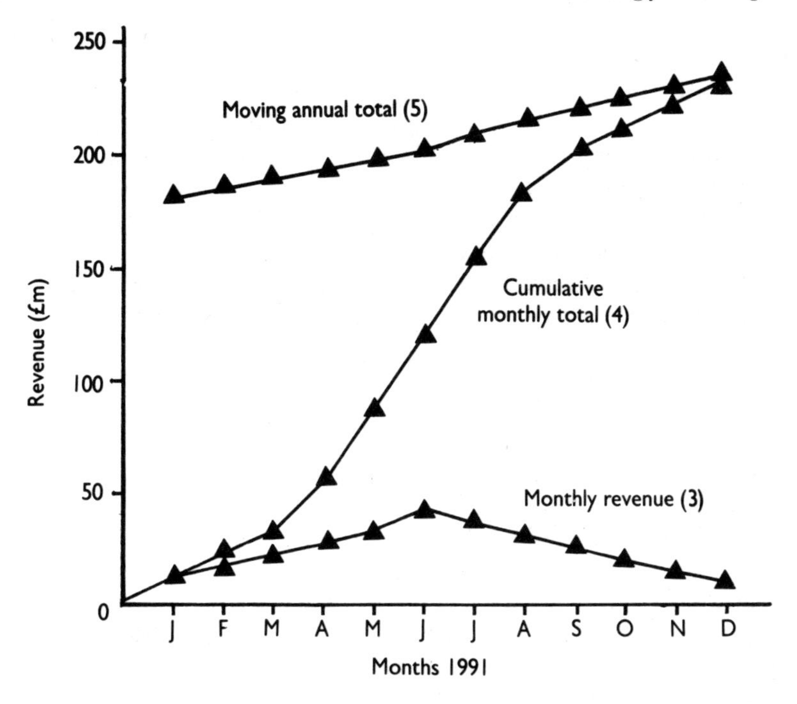

Figure 8.4 *Z chart*

Table 8.7 *Monthly revue*

(1) *Month*	(2) *Monthly revenue (£m) 1990*	(3) *Monthly revenue (£m) 1991*	(4) *Cumulative monthly total (£m) 1991*	(5) *Moving annual total (£m) 1991*
January	5	8	8	173
February	6	12	20	179
March	10	15	35	184
April	19	20	55	185
May	28	34	89	191
June	32	36	125	195
July	25	35	160	205
August	15	26	186	216
September	12	15	201	219
October	8	12	213	223
November	6	8	221	225
December	4	5	226	226
Total	170	226		

on the same axes to provide a comparison between the basic data and a trend line. The original data are plotted with the cumulative total and the moving annual total and between them the three curves form the slope of a 'Z'. This is shown in Figure 8.4 with data taken from Table 8.7. It should be noted that the figures for two years are necessary because the moving annual total is obtained by adding a new month each time and dropping a month from the previous year.

Figure 8.4 indicates that while the monthly sales fluctuate, the overall trend is upwards as shown by the moving annual total.

Looking forward

Statistics can be described as an applied form of mathematics. The management role involves decision-making, choice, planning and control. Managers receive quantities of statistical information which they need to interpret and understand and they need to collect statistical evidence when it is not available from other sources. In many cases, managers need to communicate statistical concepts to other people through reports, graphs, diagrams and presentations. The techniques outlined in this book should help managers to carry out their functions by providing an insight into statistical concepts and at least an introduction to more complex statistical methods.

The approach suggested in this book is fundamental to the use of statistics and to management decision-making and planning. A problem is identified, information is collected, this is compared with the original problem and the needs of the organisation and a solution is suggested. In arriving at a solution it is understood that this is not a 'correct' answer, it is the best answer that can be arrived at given the information, time and money available. This is the management approach and it is the statistical approach. In both cases it is understood that spending more time and money on an investigation does not necessarily lead to a much better solution. Facts and figures need to be interpreted with caution and are not a substitute for the exercise of judgement, and any forecast is surrounded by, and only as good as, the assumptions on which it is based.

Statistics provide managers with an extra tool in their workboxes, with a deeper understanding of 'facts' and an approach to their job. More sophisticated statistical techniques do not change these fundamentals; they build on them.

- *Index numbers* are measures designed to show average changes in a variable over a period of time.

$$\text{Price index} = \frac{p_1}{p_0} \times 100.$$

- A *time series* consists of numerical data recorded at intervals of time.
- A *moving average* is a process of repeatedly calculating a series of different average values along a time series in order to produce a trend line.
- *Seasonal variations* are eliminated from data in order to identify the underlying trend.
- *Linear trends* can be shown by calculating a straight line where a linear relationship exists between variables.
- A Z chart is a graphical method used to illustrate the trend against the original data.

Appendix
Basic Maths for Managers

The vocabulary of mathematics

Mathematics has its own language, with a vocabulary in the form of numbers and symbols. Like any other language, its main purpose is to enable people to exchange ideas with the minimum of effort and the maximum of clarity. Number systems form the basic vocabulary of mathematics.

The decimal system: is based on groups of ten (from the Latin decem, meaning ten). The system is based on the idea of positional value and a base of ten, so that all numerals are constructed with ten basic symbols with the actual position of any given numeral significant.

The basic symbols are 0 to 9. To represent numbers ten times as large, these digits are shifted one position to the left, the digit 0 being used as a position indicator: 10, 20, 30 ... Further increases by a factor of ten are indicated by further shifts in position: 100, 1,000, 10,000 ... A decrease by a factor of ten is shown by shifting the digit one position to the right, the digit 0 again being used when necessary to indicate position: one tenth is 0.1, one hundredth 0.01, one thousandth 0.001.

The binary system: has a base of two. The two digits used are 0 and 1 and any number can be represented by locating the digits in appropriate positions. Zero is 0, one is 1, two is 10, three is 11, four is 100, five is 101, six is 110, seven is 111, eight is 1000.

An increase in a number by a factor of two, or doubling it, is shown by shifting the digits one position to the left, 0 being used to indicate position. A halving of the number is shown by shifting the digits one position to the right. Eight is 1000, while four is 100.

Basic arithmetic

Basic operations

There are four basic arithmetical operations: addition (+), subtraction (−), multiplication (×), and division (÷).

i *addition* is the process of putting numbers together and the sign + (plus) means that the number following it is to be added to the number preceding it ($5+4=9$).

ii *subtraction* is the process of finding the difference between two numbers. The sign − (minus) indicates that the number following it is to be taken away from the number preceding it ($5-4=1$).

iii *multiplication* is the process of finding the sum of a number of quantities which are all equal to one another. It is essentially a short-cut version of adding when all the numbers are the same, so that 5×4 means $5+5+5+5=20$. The sign × (or times) indicates that the number preceding it is to be multiplied by the number following it. The result of the multiplication is called the 'product'. When two negative numbers are multiplied together they make a positive product ($-5\times-4=20$), while a positive and a negative number multiplied produce a negative product ($-5\times4=-20$).

iv *division* is the process of finding out how many times one number is contained in another number. It is indicated by the sign ÷ (or divided by) ($12\div4=3$).

The sequence of operations

Mathematical operations must be carried out in the correct sequence in order to achieve a correct mathematical calculation. Multiplication and division should be calculated before addition and subtraction; any figures in brackets should be calculated first, even before multiplication and division, and if the figures are only addition and subtraction or only multiplication and division, then the sum should be calculated by working from left to right. The mnemonic BODMAS summarises this sequence:

Brackets	First priority
Of (= multiplication) Division Multiplication	Equal priority
Addition Subtraction	Equal priority

For example:

$$8-5\times3=-7;$$
$$(8-5)\times3=9.$$

Simple arithmetic

The following sums illustrate the basic rules of arithmetic

$10+3-4\times 6$	(multiplication before addition and subtraction)	$=-11$
$5\times 6-4(7-5)$	(brackets first, multiplication next)	$=22$
$12-4+6$	(work left to right)	$=14$
$4\times 8\div 2$	(work left to right)	$=16$
-8×2	(a minus times a plus equals a minus)	$=-16$
-6×-3	(two minuses make a plus)	$=18$
$6(5)$	(a bracket means multiplication)	$=30$
$(6-2)(4+3)$	(inside the brackets first, the brackets mean multiplication)	$=28$

The basic mathematical calculations managers will be concerned with, apart from simple arithmetic, are fractions, decimals, percentages, ratios and proportions.

Fractions

Fractions are units of measurement expressed as one whole number divided by another. They allow the consideration of units of measurement smaller than a whole number. The term 'common fraction' is used to emphasise the distinction of this from the decimal fraction:

$$\frac{3}{4} \text{ as opposed to } 0.75.$$

$$\text{Common Fraction} = \frac{\text{Numerator}}{\text{Denominator}}.$$

A 'proper fraction' is one where the numerator is less than the denominator, eg:

$$\frac{3}{4}$$

while an 'improper fraction' is where the numerator is greater than the denominator, eg:

$$\frac{9}{4}.$$

Improper fractions can be reduced to a whole number and a proper fraction by dividing the numerator by the denominator, eg:

$$\frac{9}{4} \text{ becomes } 2\frac{1}{4}.$$

Fractions are added together and subtracted from one another by finding the common denominator:

$$\frac{1}{4}+\frac{5}{6}=\frac{3}{12}+\frac{10}{12}=\frac{3+10}{12}=\frac{13}{12} \quad \text{or} \quad 1\frac{1}{12}$$

$$\frac{3}{8}-\frac{1}{3}=\frac{9}{24}-\frac{8}{24}=\frac{9-8}{24}=\frac{1}{24}.$$

Fractions are multiplied together by multiplying the numerators to obtain the numerator of the answer and multiplying the denominators to obtain the denominator of the answer:

$$\frac{2}{5}\times\frac{1}{3}=\frac{2}{15}.$$

One fraction is divided by another by multiplying by its inverse:

$$\frac{2}{5}\div\frac{1}{3}=\frac{2}{5}\times\frac{3}{1}=\frac{6}{5}=1\frac{1}{5}\text{th}.$$

In order to divide mixed fractions they must first be made improper:

$$3\frac{1}{3}\div 2\frac{1}{2}=\frac{10}{3}\div\frac{5}{2}=\frac{10}{3}\times\frac{2}{5}=\frac{20}{15}=1\frac{1}{3}.$$

Decimals

A decimal number is one whose denominator is any power of

ten (10, 100, 1,000 . . .). The decimal number for a quarter:

$$\frac{1}{4} \text{ is } 0.25,$$

for a half:

$$\frac{1}{2} \text{ is } 0.5,$$

for three-quarters:

$$\frac{3}{4} \text{ is } 0.75.$$

The decimal point divides the whole number from the fraction, so that 1.25 equals the whole number plus the fraction:

$$0.25 \text{ or } \frac{1}{4}.$$

Decimal currency is based on units of ten, so that the pound sterling is divided into 100 pence and the metric system is based on units of ten:

Length:	10 millimetres (mm) = 1 centimentre (cm)
	100 centimetres = 1 metre (m)
	1,000 metres = 1 Kilometre (Km)
Capacity:	1,000 millilitres (ml) = 1 litre (l)
	1,000 litres = 1 Kilolitre (Kl)
Weight:	1,000 milligrams (mg) = 1 gram (g)
	1,000 grams = 1 Kilogram (Kg)
	1,000 Kilograms = 1 tonne (t)

The Imperial system

This system is based on a variety of units of number:

Length:	12 inches = 1 foot, 3 feet = 1 yard,
	1,760 yards = 1 mile
	(1 metre = 3 feet 3.37 inches)
Capacity:	2 pints = 1 quart, 4 quarts or 8 pints = 1 gallon
	(1 litre = 1.76 pints)
Weight:	16 ounces = 1 pound, 14 pounds = 1 stone,
	112 pounds = 1 hundredweight,
	20 hundredweight = 1 ton
	(1 Kilogram = 2.2 pounds)

Temperature

The freezing point of water is 0° in Celsius and 32° in Fahrenheit. The conversion formulae are:

$$C = \frac{5}{9}(F - 32) \quad \text{and} \quad F = \frac{9}{5}C + 32.$$

If F = 40° then:

$$C = \frac{5}{9}(40 - 32) = 4.4°\text{C}.$$

If C = 20° then:

$$F = \frac{9}{5} \times 20 + 32 = 68°\text{F}.$$

Percentages

Per cent (%) means per hundred, so that 40 per cent is 40 per hundred or 40 out of a hundred. Percentages are arrived at as follows:

$$\frac{3}{4} = 75\%: \quad \frac{3}{4} \times 100$$

$$0.5 = 50\%: \quad 0.5 \times 100$$

$$75\% = \frac{3}{4}: \quad \frac{75}{100}$$

$$75\% = 0.75: \quad \frac{75}{100}.$$

$$\text{£10 as a percentage of £50} = \frac{10}{50} \times 100 = 20\%$$

$$10\% \text{ of £25} = \frac{10}{100} \times 25 = \text{£2.50}.$$

Powers and square roots

'Powers' or exponents are a shorthand method of representing multiplications.

$$6^3 = 6 \times 6 \times 6 = 216 \quad (6 \times 6 = 36 \times 6 = 216)$$

This is 6 to the power 3.

The square root of a number is that number which when multiplied by itself gives the original number. The square root of 36 is 6 because $6 \times 6 = 36$. This is expressed as $\sqrt{36} = 6$.

Ratios

A ratio is a relationship between the quantities expressed in a number of units which enables comparison to be made between them.

For example, the direct costs of a firm may consist of 80 per cent wages and salaries and 20 per cent raw materials, power and transport. The ratio of labour costs to other costs can be said to be 80% : 20% or 4 : 1. This means that for every unit of 'other' costs there are 4 units of labour costs.

Proportions

This is the amount a part of a unit makes up of the whole.

For example, if a firm's labour costs are £4 million and the total costs are £5 million, then labour costs make up 80 per cent of total costs or:

$$\frac{4}{5}\text{ths.}$$

Interest rates

Simple interest is based on an arithmetic progression, while compound interest is based on a geometric progression.

Simple interest: £100 invested for five years at a simple interest rate of 10 per cent per annum, at the end of the five years the total amount accumulated would be £150:

$$£100 + £10 + £10 + £10 + £10 + £10 = £150.$$

The formula is:

$$A = P(1+tr)$$

where A is the total amount accumulated

P is the original investment

t is the time in years

r is the rate of interest

$$A = £100\left(1+5\times\frac{10}{100}\right) = £100\times 1.5 = £150.$$

Compound interest: £100 invested for five years at a compound interest rate of 10 per cent per annum, at the end of five years the total amount accumulated would be £161.05:

original investment = £100

Year 1 = £110 (£100 + 10%)
Year 2 = £121 (£110 + 10%)
Year 3 = £133.1 (£121 + 10%)
Year 4 = £146.41 (£133.1 + 10%)
Year 5 = £161.05 (£146.41 + 10%).

The formula is:

$$A = P(1+r)^t$$

where A is the total amount accumulated

P is the original investment

r is the rate of interest

t is the time in years

$$A = £100\left(1+\frac{10}{100}\right)^5 = £100\times 1.6105 = £161.05.$$

Present value

This is a concept which is the opposite of compound interest in that it answers questions about how much money to invest to achieve a particular return given a rate of interest and a period of time.

The formula is:

$$P = \frac{A}{(1+r)^t}$$

where P is the original investment or the 'present value'

A is the total amount accumulated

r is the rate of interest

t is the time in years.

If the objective is to achieve a return of £100 in four years time, given a rate of interest of 5 per cent, then:

$$P = \frac{100}{\left(1+\frac{5}{100}\right)^4}$$

$$= \frac{100}{1.2155} = £82.27.$$

This means that an investment of £82.27 at an interest rate of 5 per cent over four years will achieve a return of £100.

If the interest rate had been 10 per cent the present value (the amount to be invested) would have been £68.30.

Discounted cash flow

This involves the calculation of the present value of a series of future cash flows. By calculating present values or the net present value of a proposed investment project a business can ascertain its present worth and faced with a choice between alternative investment a business can calculate their net present values to provide a means of comparison.

In the calculation of present values, the amounts to be invested to achieve a return of £100 were £82.27 and £68.30. Clearly the latter would be a more profitable investment. Over a period of years this may depend on the interest or discount rate, on the original cost of investment and on the estimated annual cash flows. As in the above example, the discount rate (or the discounted rate of return) may play a vital part in any decision.

Levels of measurement

Different scales are used for particular levels of measurement in mathematics. These are shown in the box below.

Levels of Measurement

Scales	*Definition*	*Characteristics**
Nominal	classification labelling	symmetry transitivity
Numbers = names, one number is not greater or better than another. Example: house numbers.		
Ordinal	order rank	symmetry transitivity asymmetrical
Numbers are ranked as greater or smaller than each other without being able to say how much greater or smaller. Example: a toaster may have numbers to show scales of brownness.		
Interval	value	symmetry transitivity asymmetrical units of measurement
Numbers are ranked as greater or smaller than each other and the exact distances between them are indicated. Example: age, height, weight are all exact measurements.		

* See definitions which follow the box.

Definitions

Symmetry: a relationship between *A* and *B* also is true between *B* and *A*. If house No. 3 is opposite house No. 8, then No. 8 is opposite No. 3.

Transitivity: if A = *B* and *B* = C, then *A* = C. If house No. 3 is the same style as house No. 8, and No. 8 is the same style as No. 4, then No. 3 is the same style as No. 4.

Asymmetrical: a special relationship may hold between *A* and *B* which does not hold between *B* and *A*. If house No. 5 is larger than house No. 9, then No. 9 cannot be larger than No. 5.

Units of measurement: physical units agreed upon as a common standard which can be applied repeatedly with the same results, such as the metric system. The number 10 is twice 5, the difference between 35 and 50 is 15.

Accuracy

Levels of tolerance

There are times when managers need to be very precise in their calculations, while at other times approximations may be sufficient. In mathematics perfect accuracy is possible and calculations can lead to the 'right' answer. In statistical applications this is far less likely to be the case and neither will it be necessary. In so far as statistics are the starting point for investigations or are used to narrow the field of discussion, there will be levels of accuracy that can be accepted.

Levels of tolerance are concerned with the degree of accuracy required in particular circumstances. In measuring component parts of an engine, a high level of accuracy may be required with very narrow limits of tolerance. In measuring the length of building nails there may be wider limits of tolerance because it is not crucial for them to be 'exactly' the same length.

Error

In statistical terms, error is the difference between the true figure (or what can be accepted as this) and what is taken for an estimate or approximation.

Rounding

Figures are often rounded to remove fractions or decimals. The usual procedure is to 'round to the nearest whole number' with 0.5 and above rounded up and under 0.5 rounded down.

Another method used to round figures is to remove or 'truncate' unwanted digits (1.9423 would become 1.9 or 1.94). The use of 'significant figures' serves the same purpose: 1.9423 to four significant figures would be 1.942, to three significant figures 1.94 and to two significant figures 1.9.

Absolute and relative error

Absolute error is the difference between an approximation or estimate and the true figure, whereas the relative error is the absolute error divided by the estimate. For example, if a manager pays an employee £140 when the payment should have been £150, the relative error is about 7 per cent:

$$\left(\frac{10}{140} \times 100\right)$$

while the absolute error is £10. If a manager pays another employee £220 when the payment should have been £230, the absolute error is the same as before at £10. The relative error is:

$$\frac{10}{220} \times 100$$

or about 4.5 per cent.

Symbols of Mathematics

The main mathematical symbols used in statistics are:

x:	a collective symbol meaning all the individual values of a variable
y:	an alternative symbol to x
$\bar{x}$:	the arithmetic mean
n:	the number of items in a collection of figures
f (frequency):	the number of times a given value occurs in a collection of figures
Σ (sigma):	the sum of
d (deviation):	the difference between the values
$=$:	equals
$\simeq$:	approximately equals
$\neq$:	not equal to
$>$:	greater than, larger than, more than
$<$:	smaller than, less than
σ (small sigma):	standard deviation
r:	coefficient of correlation
r':	coefficient of rank correlation
μ (mew or mu):	the population mean in sampling

Index